Dynamic System Reconfiguration in Heterogeneous Platforms

Lecture Notes in Electrical Engineering

Volume 40

For other titles published in this series, go to
www.springer.com/series/7818

Nikolaos S. Voros • Alberto Rosti
Michael Hübner
Editors

Dynamic System Reconfiguration in Heterogeneous Platforms

The MORPHEUS Approach

Springer

Editors
Nikolaos S. Voros
Technological Educational
Institute of Messolonghi
Department of Telecommunication
Systems & Networks
Greece

Alberto Rosti
STMicroelectronics
Italy

Michael Hübner
ITIV
University of Karlsruhe (TH)
Germany

ISBN: 978-90-481-2426-8 e-ISBN: 978-90-481-2427-5
DOI: 10.1007/978-90-481-2427-5
Springer Dordrecht Heidelberg London New York

Library of Congress Control Number: 2009926321

Cover design: eStudioCalamar Figueres, Berlin

Printed on acid-free paper

Springer is part of Springer Science+Business Media (www.springer.com)

Preface

Arthur Schopenhauer: "Approximately every 30 years, we declare the scientific, literary and artistic spirit of the age bankrupt. In time, the accumulation of errors collapses under the absurdity of its own weight."

Reiner Hartenstein: "Mesmerized by the Gordon Moore Curve, we in CS slowed down our learning curve. Finally, after 60 years, we are witnessing the spirit from the Mainframe Age collapsing under the von Neumann Syndrome."

This book intends to serve as a basis for training on reconfigurable architecture. All contributions to the book have been carefully written, focusing on the pedagogical aspect so as to become a relevant teaching material. Therefore, this book addresses in particular students, postgraduate programmers/engineers or anyone interested to learn about dynamic system reconfiguration in heterogeneous platforms through the European MORPHEUS project. This preface also introduces to the historical background and the significance of Reconfigurable Computing, and highlights the impact of MORPHEUS and its key issues.

About the History of Reconfigurable Computing

Since 400 years ago Galileo Galilei started using a telescope for astronomical observations, the International Astronomic Union in Paris kicked off the International Year of Astronomy 2009. Galilei's observations contributed to the revolution which finally by Nikolaus Kopernikus smashed the geo-centric Aristotelian flat world model. By this occasion we should be aware, that also the CPU-centric world model taught by mainstream CS education is as flat as the antique geo-centric world. Going far beyond navel-gazing onto this excellent project we can recognize that this book and the MORPHEUS environment are part of a world-wide run-away counter-revolution. Compared to the astronomical world model the history of the basic computing world model went backwards. The Kopernikan model came first. In 1884 Hermann Hollerith finished the prototype of the first electrical computer: data-stream-driven and reconfigurable. This has happened exactly 100 years before Xilinx introduced the first FPGA: data-stream-driven by very fast interfaces instead of punched cards and reconfigurable by modern technology instead of a bulky plug board.

A Wrong Decision with Serious Consequences

Roughly 60 years later with ENIAC the US army triggered a backward paradigm shift, over to the instruction-stream-based hardwired von Neumann model – just for creating ballistic tables. This means that the basic mind set of mainstream computing science and practice started going backward by moving to the dominance of the CPU-centric quasi Aristotelian flat world model: from Kopernikus back to Aristoteles. Meanwhile we have learnt, that this shift has been a wrong decision with serious consequences. This has been also the main motivation to set up the MORPHEUS project: The limitations of conventional processors are becoming more and more evident. Even a 4 or 5 GHz CPU in a PC to-day does not even have the performance to drive its own display. As co-processors we need a variety of accelerators which we call ASICs (Application-Specific ICs): the tail is wagging the dog. The MORPHEUS system is a major breakthrough leading us away from the CPU-centric Aristotelian flat world model of computing from the mainframe age, disruptively forward to the twin-paradigm Kopernikan modern world image of computing, also supporting the replacement of ASICS by FPGAs, etc.

Data Processing Abolished

Directly run by data sequencing the Hollerith machine has been directly a data processing machine. It has been a data stream machine. We should not get confused by looking at sequencing punched cards. We should always be aware of the Hollerith machine being a very simple forerunner of Reconfigurable Computing (RC). Its replacement by the von Neumann machine means the abolishment of direct data processing. The von Neumann machine is an instruction processing machine. Its sequencing is controlled by the program counter. Running data streams, i.e. the movement and addressing of data has emigrated into the software domain, i.e. from the machine level into the application. Being fully instruction-stream-centric the basic machine architecture model has turned from Kopernikan into Aristotelian. Together with the memory wall this is a reason of massive inefficiency. Already in the 1980s we have published a speed-up by up to a factor of 15,000 by using reconfigurable address generators and data counters instead of a program counter. It is an important merit of MORPHEUS to find a well practicable solution to cope with these efficiency problems (also see below for the section "Mastering memory requirements.")

The Decline of a Growth Industry

As chips get smaller and smaller, they grow intensely hot, power-hungry and unreliable. After four decades the "free ride on Gordon Moore's Law" has reached its end by stopping the GigaCycle clock speed race, because of cooling problems (up to almost 200 W per processor) – in favor of an increasing number of lower power processor cores on a single chip: maybe increasing by a factor of 2 every 2 years or slightly faster. This leads to the many-core programming crisis, since here a methodology and a sufficiently large programmer population qualified for this kind of parallel programming, also onto hetero solutions, is far from being existent, and, the industry's decline from growth industry to replacement business is looming. Semiconductor technology currently stalls at the "20 nanometer wall" (manufacturability questioned) and is going to come to a final halt definitely in about ten years around 10 nanometers, where chips will massively lose their efficiency, being ruled by the laws of quantum mechanics.

Our Electricity Bill May Get Unaffordable

More and more, data centers and supercomputer centers complaint about their rapidly growing huge electricity bills. The internet, a major energy guzzler, is growing: more bandwidth, more servers, more users. The electricity consumption of its server farms is growing 20% per year. Together with all communication networks it takes 3% and after 23 years it will reach 100% of today's total electricity consumption. Extending the IT state of the art world-wide would require 40% of today's power plants, and in less than 10 years 100% would be exceeded. However, the internet is only a minor part of our computer-based infrastructures. This is also an important political aspect as (1) an enormous cost factor for our economy, which might become unaffordable in the future, (2) a weighty climate pollution factor, and (3) for securing our energy supply, in Europe especially important because of depending on crisis-prone suppliers and embargo threats. Several studies forewarn new energy price explosions by shortages and cartels. For these reasons a drastic reduction of our computing-related electricity consumption is very important and is effectively possible by Reconfigurable Computing (RC).

Massively Saving Energy

More recent developments in industry and academia teach us, that this – still mainstream – instruction-stream-based von-Neumann-only mind set of this CPU-centric world is running out of steam soon. From the scene of embedded computing (non-von Neumann) reconfigurable platforms are proceeding from niche to mainstream. Let's illustrate their energy efficiency. "The human brain needs just a few watts for about 10,000,000,000,000,000 (10 Million Billion) operation steps per second", said Alfred Fuchs: "based on contemporary von Neumann type computing our brain would need more than a Megawatt, the electricity consumption of a small town: our brain would burn up within fractions of a second." According to another estimation several nuclear power stations would be needed to run a real-time simulation of the brain of a mouse. What do we learn from this wonder of nature? The consequence is an adaptive microchip architecture of masses of cells massively working in parallel. Such a technology is already existing. By experts called "FPGA" (Field-Programmable Gate Array) it's celebrating fascinating success on the multi-billion dollar market of Reconfigurable Computing (RC): as a massively energy saving micro brain being dramatically more powerful than a CPU.

The von Neumann Syndrome

Why is this illustration so interesting for us? For politics, economy, and climate protection it is time to recognize the immense energy consumption facts for running all visible and embedded von Neumann computers in the world, also including computer-based cyber infrastructures. Decades ago the von Neumann model has been criticized by celebrities like Dijkstra, Backus and Arvind. Nathan's law, also attributed to Bill Gates, models software as a gas which completely fills up any available storage space at all levels. By the term "von Neumann Syndrome" Prof. C. V. Ramamoorthy from UC Berkeley summarized the exploding code complexity yielding software packages of up to astronomical dimensions, and interminable notorious abundance of management and quality problems, security problems, and many other problems which are typical to our CPU-centric flat world model. But we cannot relinquish the services we obtain from it. We should teach funding agencies and politicians, also though mass media, that a highly effective technology is available for the way out of this dilemma: Reconfigurable Computing (RC).

Merging Three Competing Solutions

The use of RC reminds us, that to implement architectures for demanding computation problems of today there are basically three competing solutions: high-performance microprocessor CPUs, RC platforms, and ASICs. Often high performance processor CPUs are not sufficiently efficient for certain applications and some of them may have a power consumption up to 200 W. ASICs are unaffordable for low market volume products. Between ASICs and microprocessors, RC is rapidly gaining momentum as a third solution becoming more and more attractive for applications such as cryptography, streaming video, image processing and floating point operations. The limitations of conventional CPUs are becoming more and more evident. The growing importance of stream-based applications makes coarse-grain reconfigurable architectures an attractive alternative. They combine the performance of ASICs with the flexibility of CPUs. On the other hand, irregular control-flow dominated algorithms require high-performance sequential processor kernels for embedded applications.

Speed-Up and Saving Energy by Migration to Configware

Reconfigurable computing solutions are by up to orders of magnitude more powerful (up to a factor of 34,000 for breaking DES encryptions [GWU]) and drastically energy saving (down to less than a thousandth [GWU]) alternative to the von Neumann model and its programming. Such migrations promise improvements by up to several orders of magnitude, whereas "green computing" and low power design methods provide much less benefit than one order of magnitude. For migrations the improvement depends on the type of algorithm. For instance, DSP, multimedia, image processing, and bio-informatics applications yield excellent results. The other extreme are error-correcting decoding algorithms for wireless communication, which require an enormous amount of wiring resources when mapped from time to space. Also applications with many irregularly structured control parts and other spaghetti or sauerkraut structured programs are bad candidates. So we need a good taxonomy of algorithms to decide, which applications should be migrated first. It will not make sense, to migrate all kinds of algorithms, so that we will always have a twin-paradigm world with co-existence of both, CPUs and RC platforms.

Mastering the Challenge by Twin Paradigm

The requirements of high performance embedded systems raise a grand scientific and technical challenge which can be met by such an integration of software and hardware. This twin paradigm approach combining the instruction-stream-based mind set of software with the data-stream-based mind set of hardware people is successfully demonstrated by the MOPHEUS project. Within such a spatial/temporal computation twin-paradigm approach reconfigurable computing the MORPHEUS methodology supports finding a trade-off between computation in time (instruction-stream-driven by software: slower, but with smaller resources) and spatial computation (kind of data-stream-driven, programmed by configware: fast but with large resources). Also outside the embedded systems scene we need such a paradigm shift. Its flexibility given by a clever choice of CPU cores and by fine grain, medium grain, and coarse grain reconfigurable modules (supporting stream processing mapped onto execution pipelines across multiple HREs), as well as by tools supporting a well organized design flow, brings it very close to a general purpose twin paradigm platform. MORPHEUS combines the most promising approaches to post-fabrication customizable hardware in an integrated platform concept. The heterogeneous set of programmable computation units adapt to a broad class of applications. Massively superior to fine-grained-only reconfigurable solutions MORPHEUS is an important trailblazer of the counter-revolution leading us from a CPU-centric flat world to a fascinating Kopernican computing universe.

Mastering Memory Requirements

Satisfying such memory requirements is no easy task, and SDRAM interfaces have long been a critical performance bottleneck. However, by taking advantage of the MORPHEUS memory access optimization method, these limitations can be greatly reduced. In the MORPHEUS project, a bandwidth-optimized custom

DDR-SDRAM memory controller meets the massive external memory requirements of each of the planned applications, not met by off-the-shelf memory controllers. High-end multiple processing applications with demanding memory bandwidth requirements, implemented on MORPHEUS fully demonstrate its potential as a high-performance reconfigurable architecture.

A New Class of Reconfigurable Platforms

After the success story of statically reconfigurable FPGAs again a new class of reconfigurable platforms is emerging since more than a decade: dynamically reconfigurable multi-core architectures (like MORPHEUS) able to efficiently cope with rapidly changing requirements. Dynamic reconfiguration allows changing the hardware configuration during the execution of tasks. With devices that have the capability of run time reconfiguration (RTR), multitasking is possible and very high silicon reusability can be achieved. A reason to use run time scheduling techniques is s growing class of embedded systems which need to execute multiple applications concurrently with highly dynamic behavior. Turing award recipient Joseph Sifakis has summarized in his keynote at WSEAS CSCC 2008, that designing embedded systems requires techniques taking into account extra-functional often critical requirements regarding optimal use of resources such as time, memory and energy while ensuring autonomy, reactivity and robustness.

The MORPHEUS Architecture

The architecture of the MORPHEUS platform is essentially made of techniques available from industry: a control processor and multi-layer AMBA busses also including DMA and a control processor (from ARM), three heterogeneous, reconfigurable processing engines (HREs), a memory hierarchy and common interface for accelerators and an efficient and (from ST's spidergon technology) scalable communication and configuration system (NOC, DNA, PCM) based on (from ST's spidergon technology) provides a routing mechanism. The accelerator HREs within MORPHEUS are: a coarse-grained reconfigurable array (XPP from PACT) for high bandwidth data streaming applications, the medium-grained DREAM reconfigurable array (PiCoGA core from ARCES), and FlexEOS the fine-grain reconfigurable M2000, an embedded FPGA (eFPGA).

A Significant Breakthrough

Reconfigurable architectures like MORPHEUS represent a significant breakthrough in the embedded systems research. The ambition of the MORPHEUS reconfigurable platform is to deliver processing power competitive with state-of-the-art

Systems-On-Chip, while maintaining high flexibility to a broad spectrum of application, and user-friendliness. The ambition has been implemented successfully.

Trailblazing MORPHEUS Demo Applications

The MORPHEUS platform's potential is demonstrated for several application domains including reconfigurable broadband wireless access and network routing systems, processing for intelligent cameras used in security applications, and film grain noise reduction for use in high definition video. Unlike some other applications, the image-based applications have been shown to exhibit immense memory needs. Due to high data rates real-time post processing for film grain noise reduction in digital cinema movies is extremely challenging and beyond the scope of standard DSP processors (and ASICs have to a small market volume). Here post processing requiring up to 2,000 operations per each of 3 million pixels results in high memory and computational needs that necessitate accelerators, such, that dedicated hardware usually is unaffordable. A proven answer is the MORPHEUS platform with mixed granularity reconfigurable processing engines and an integrated toolset for rapid application development.

MORPHEUS Application Development makes use of a number of successful tools from industry. For instance, the coarse-grained XPP array is programmable in C and includes a cycle-accurate simulator and a complete development environment, and, instruction level parallelism for the medium-grained DREAM reconfigurable array can be automatically extracted from a C-subset language called Griffy-C.The first objective is to hide the HREs heterogeneity and abstract the hardware details for the programmer. A Control Data Flow Graph (CDFG) format is used as an intermediate and technology independent format An innovative aspect of the MORPHEUS approach is the seamless design flow from a high level description toward target executable code. An ambition of the MORPHEUS toolset is to abstract the heterogeneity and the complexity of the architecture in such a way that software designers are able to program it (without knowledge and experience of HREs). With run time reconfiguration (RTR) multitasking is possible and very high silicon reusability can be achieved. In MORPHEUS this is managed by the Predictive Configuration Manager (PCM), which basically hides the context switching overhead and abstracts the interface of the reconfigurable engine from the designer's point of view by an intermediate graphical representation of the applications extracted at design time.

MORPHEUS: Close to Being General Purpose

The single chip MORPHEUS is close to a large "general purpose" multi-core platform (homogeneous and heterogeneous) for intensive digital signal processing, By embedded dynamically reconfigurable computing completed by a software (SW) oriented design flow. These "Soft Hardware" architectures will enable huge computing density improvements (GOPS/Watt) by a factor of 5 compared to FPGAs, flexibility and improved time to market thanks to a convenient programming toolset. In comparison with FPGA-based systems like Stratix or Virtex, MORPHEUS can offer a larger computation density thanks to the innovative contribution of fine/medium/coarse-grained programmable units.

Establishing the European Foundation

MORPHEUS ambitions are to establish the European foundation for a new concept of flexible"domain focused platforms", positioned between general purpose flexible HW and von Neumann general purpose processors. The major project results are (a) a modular silicon demonstrator composed of complementary run-time reconfigurable building blocks to address the different types of application requirements and (b) the corresponding integrated design flow supporting the fast exploration of hardware and software alternatives. The suitability and the efficiency of the proposed approach have been validated by a set of complementary test implementation cases which include: a real-time digital film processing system, an Ethernet based in-service reconfiguration of SoCs in telecommunication networks, a homeland Security – Image processing system for intelligent cameras and a system implementing the physical layer of 802.16j mobile wireless system.

MORPHEUS Model of Changing Market Trends

Solutions to meet education needs and to narrow the designer/programmer productivity gap are offered by the MORPHEUS training methods. MORPHEUS platforms along with the seamless MORPHEUS tool design flow would be the ideal resource to organize the twin-paradigm lab courses for the software/configware co-education urgently needed. We need Une Levée en Masses: professors back to school! To maintain the growth rate we are used to, we need a lot of software to configware migrations as well as a sufficiently large new breed of programmer population qualified to do that. The emerging new computing landscape will also affect market trends. The share of "RC inside" will massively grow, although some "CPU inside" will still be needed. We have to expect also an impact on the EDA market by new kinds of design flows. The MORPHEUS design system is an excellent prototype giving us the vision.

Fascinating New Horizons appear from market trends, from the vision of the MORPHEUS project and its highly convincing deliverables. MORPHEUS provides a highly attractive platform chip and board, along with a very user-friendly application development framework and training methods reported within this book. However, we have to take care, that a sufficiently large qualified programmer population is available for a world-wide massive break-through. The Karlsruhe

Institute of Technology and the University of Brasilia are cooperating to come up with an innovative text book for dual rail education to overcome the "traditional" software/hardware chasm and the software/configware chasm by a twin dichotomy approach with: (1) the paradigm dichotomy, and (2) the relativity dichotomy where (1) the paradigm dichotomy provides a twin paradigm model connecting the von Neumann machine (with program counter) with the datastream machine (using data counters), and, (2) the relativity dichotomy model provides mapping rules between the time domain and the space domain, for instance to support parallelization by time to space mapping.

Putting Old Ideas Into Practice (POIIP)

Such innovative twin-dichotomy introductory teaching methods are intuitive by the fact that their imperative programming languages have exactly the same language primitives with a single exception. Data stream languages feature parallelism within loops, whereas instruction stream languages do not. David Parnas said, that "The biggest payoff will come from Putting Old ideas into Practice and teaching people how to apply them properly." This also holds for mapping between time and space based on simple key rules of thumb: (1) loop turns into pipeline (software to hardware and software to configware migration), and, (2) a decision box from the program flow chart turns into demultiplexer.

Continued Growth Beyond Moore's Law

Growth rates we are used to, from the free ride on Gordon Moore's Law, can be continued for at least two more decades by a mass migration campaign for migrating selected applications from software to configware. This is possible as soon as solutions of the education dilemma have been started. Since such movements take some effort and a lot of time this is also a chance to create a lot of jobs for the next two decades. The benefit from massively saving energy we will not be obtained without effort. However, what is urgently needed to cope with the manycore crisis is the "Configware Revolution": a world-wide disruptive education reform leading to a far-reaching software/configware co-education strategy.

The Configware Revolution

Comparable to the current programmer population's qualification gap w.r.t manycore programming combined with configware programming has been the scenario around 1980, where a designer population qualified to cope with thousands of transistors on a microchip has not been existing. This has been the VLSI design crisis – the missing reply to Moore's Law. This was the reason of the VLSI Design Revolution, brain child of Carver Mead Lynn Conway, a world-wide Levée en

Masses creating the missing designer population and being the incubator of the EDA industry: the most influential research project in modern computer history. Now, after about 30 years and inspired by the MORPHEUS methodology and its training methods, we need a similar influential far-reaching revolution: the Configware Revolution.

Reiner Hartenstein
http://hartenstein.de

Acknowledgments

The research work that provided the material for this book was carried out during 2005–2008 mainly in the MORPHEUS Integrated Project (Multi-purpose dynamically Reconfigurable Platform for intensive Heterogeneous processing) supported partially by the European Commission under the contract number 027342. Guidance and comments of EU Project Officers and EU Reviewers on research direction have been highly appreciated.

In addition to the authors, the management teams of the partners participating in the MORPHEUS consortium are gratefully acknowledged for their valuable support and for their role in the project initiative.

The editors express their special thanks to Mr. Gilbert Edelin, Prof. Jürgen Becker and Prof. Reiner Hartenstein for their valuable remarks and their willingness to contribute to the final review of the material presented in the book.

Contents

Contributors

Cyrille Batariere
Thales Optronics S.A., France

Jürgen Becker
ITIV, University of Karlsruhe (TH), Germany

Marcel Beemster
ACE BV, The Netherlands

Koen Bertels
Delft University of Technology, The Netherlands

Philippe Bonnot
Thales Research & Technology, France

Paul Brelet
Thales Research & Technology, France

Fabio Campi
STMicroelectronics, Italy

Stéphane Chevobbe
CEA LIST, France

Luca Ciccarelli
STMicroelectronics, Italy

Marcello Coppola
STMicroelectronics, France

Antonio Deledda
ARCES – University of Bologna, Italy

Rolf Ernst
IDA, TU Braunschweig, Germany

Sebastian Goller
Chemnitz University of Technology, Germany

Arnaud Grasset
Thales Research & Technology, France

Hélène Gros
ARTTIC SAS, France

Stéphane Guyetant
CEA LIST, France

Ulrich Heinkel
Chemnitz University of Technology, Germany

Michael Hübner
ITIV, University of Karlsruhe (TH), Germany

David Hulance
M2000, France

Frank Ieromnimon
Intracom Telecom Solutions S.A., Greece

Massimiliano Innocenti
STMicroelectronics, Italy

Joachim Knaeblein
Alcatel-Lucent, Nuremberg, Germany

Matthias Kühnle
ITIV, University of Karlsruhe (TH), Germany

Loic Lagadec
Université de Bretagne Occidentale, France

Riccardo Locatelli
STMicroelectronics, France

Tommaso De Marco
ARCES – University of Bologna, Italy

Erik Markert
Chemnitz University of Technology, Germany

Giuseppe Maruccia
STMicroelectronics, France

Philippe Millet
Thales Research & Technology, France

Claudio Mucci
STMicroelectronics, Italy

Sophie Oriol
ARTTIC SAS, France

Elena Moscu Panainte
Delft University of Technology, The Netherlands

Stylianos Perissakis
Intracom Telecom Solutions S.A., Greece

Luca Perugini
STMicroelectronics, Italy

Damien Picard
Université de Bretagne Occidentale, France

Lorenzo Pieralisi
STMicroelectronics, France

Matteo Pizzotti
STMicroelectronics, Italy

Bernard Pottier
Université de Bretagne Occidentale, France

Gabriele Pulini
M2000, France

Wolfram Putzke-Röming
Deutsche THOMSON OHG, Germany

Florian Ries
ARCES – University of Bologna, Italy

Davide Rossi
ARCES – University of Bologna, Italy

Alberto Rosti
STMicroelectronics, Italy

Henning Sahlbach
IDA, TU Braunschweig, Germany

Axel Schneider
Alcatel-Lucent, Nuremberg, Germany

Marius Schoorel
ACE BV, The Netherlands

Eberhard Schüler
PACT XPP Technologies, Germany

Vlad-Mihai Sima
Delft University of Technology, The Netherlands

Graham Stephen
CriticalBlue Ltd, United Kingdom

Richard Taylor
CriticalBlue Ltd, United Kingdom

Florian Thoma
ITIV, University of Karlsruhe (TH), Germany

Luca Vanzolini
STMicroelectronics, Italy

Nikolaos S. Voros
Technological Educational Institute of Mesolonghi, Greece

Markus Weinhardt
PACT XPP Technologies, Germany

Sean Whitty
IDA, TU Braunschweig, Germany

Part I
Introduction to MORPHEUS

Chapter 1
Introduction

A Heterogeneous Dynamically Reconfigurable SoC

Philippe Bonnot, Alberto Rosti, Fabio Campi, Wolfram Putzke-Röming, Nikolaos S. Voros, Michael Hübner, and Hélène Gros

Abstract The objectives of high performance and low global cost for embedded systems motivate the approach that is presented here. This approach aims at taking benefit of reconfigurable computing implemented on System-on-Chip (SoC) including host processors. The proposed architecture is heterogeneous, involving different kinds of reconfigurable technologies. Several mechanisms are offered to simplify the utilization of these reconfigurable accelerators dynamically. The approach includes a toolset that permits a software-like methodology for the implementation of applications. The principles and corresponding realizations have been developed within the MORPHEUS project, co-funded by the European Union in the sixth R&D Framework Program.

Keywords Reconfigurable computing · SoC · heterogeneous architectures · dynamic reconfiguration · toolset · embedded systems · European Union · EU-funded project · FP6 program · collaborative project

---

P. Bonnot (✉)
Thales Research & Technology, France
philippe.bonnot@thalesgroup.com

A. Rosti and F. Campi
STMicroelectronics, Italy

W. Putzke-Röming
Deutsche THOMSON OHG, Germany

N.S. Voros
Technological Educational Institute of Mesolonghi (consultant to Intracom Telecom Solutions S.A.), Greece

M. Hübner
ITIV, University of Karlsruhe (TH), Germany

H. Gros
ARTTIC SAS, France

N. Voros et al. (eds.), *Dynamic System Reconfiguration in Heterogeneous Platforms*, Lecture Notes in Electrical Engineering 40,

1.1 Motivations for an Heterogeneous Reconfigurable SoC

From embedded systems industry perspective, and especially for the applications specifically considered here (wireless telecommunication, video film processing, networking and smart cameras), cost effectiveness is essential. This means that the processing components used in such systems must provide high density performance and enable inexpensive development.

The main characteristics of the selected application systems can already be identified here before a deeper presentation in further chapters. These systems belong to the domain of data flow driven streaming application. They include a significant part of intensive and regular processing well suited for hardware-like implementation. These more and more intelligent systems also require a flexible platform: it is not possible to merely use hardwired functions because the system has to adapt itself to external constraints. Moreover, these constraints possibly vary in time, either because of an evolution of the mission or for efficiency optimization reason. Another important characteristic lies in the complexity of these systems: They are made of many heterogeneous functions, they handle complex data structures, etc. Some efficient tools are therefore required to manage this complexity and enable a competitive time-to-market. Especially the handling of complexity is essential in this context. The design space provided by such heterogeneous system is hardly or not exploitable from a developer without the tool support.

The approach that has been consequently identified, and that will be described in this book, aims at taking benefit of the cumulated advantages of reconfigurable technologies, General-Purpose Processors and SoC technique to satisfy the requirements mentioned above.

Reconfigurable technologies have indeed several advantages. These advantages are the performance density coming from the capability to reuse the same hardware for several functions (flexibility) and also from the efficient dataflow programming model which is provided (for the applications where this model can apply). It is clear that an implementation solution purely based on the von Neumann model could not be appropriate to the kind of applications that is targeted here. Moreover, reconfigurable technologies permit customized implementations for specific functions (for instance, specific data formats). The various existing types of reconfigurable technology (fine grain, coarse grain) also permit to choose the most suited solution for a targeted function. However, these types of devices are generally difficult to program and require specific skills. This difficulty has its background also from the huge design space.

The advantages of General Purpose Processors are also significant. They notably bring a high flexibility which allows a very easy control of a system. They can easily be programmed.

SoC technique is a way to perform the integration of selected hardware modules (also named IP) on a chip. This is an efficient way to integrate various types of reconfigurable accelerators, which can be chosen for their complementarities or because they are well adapted to specific requirements. In the strategy presented here, the reconfigurable units build a heterogeneous system around the various types of accelerators that are generally useful for the type of applications targeted here.

Thanks to the various types of reconfigurable technology that can thus be integrated in such SoC, one can take benefit of different types of accelerators. A first type best suits for arithmetic computation such as FFT, DCT and real-time image processing for instance. Another type can better handle bit-level computation, finite state machines for control dominant functions. Also some other type (in between those two extreme cases) suits for mixed functions ranging from error correction to binarized image processing.

The presented approach is therefore based on a SoC where accelerators can take benefit of the performance advantages of reconfigurability (and its various types) and also benefit of the programmability of general purpose processor that controls this SoC. One main purpose of this book is to explain how the combination of these elements allows building a competitive processing solution. This includes the definition of an efficient programming solution which is a key point required to efficiently develop products from this SoC. This aspect is a crucial part of the presented approach. The reasons mentioned above motivated the MORPHEUS consortium to define and develop this approach in the frame of the MORPHEUS European Project co-funded by the European Union in the frame of the sixth R&D Framework Program. MORPHEUS stands for "Multi-purpose dynamically Reconfigurable Platform for intensive Heterogeneous processing". The name of the project is a reference to the ability of this god from ancient Greek mythology to take any human's form and appear in their dreams.

1.2 Overview of the MORPHEUS Concept

The MORPHEUS concept has been defined to offer a platform that can take benefit from the performance density advantage of reconfigurable technologies and the easy control capabilities of general purpose processors, implemented by a SoC as explained above and summarized in Fig. 1.1. The concept consists of a SoC with the following characteristics:

- A regular system infrastructure (communications and memories) hosting heterogeneous reconfigurable accelerators
- Dynamic reconfiguration capabilities
- Data-stream management capabilities
- A software-oriented approach for implementation

The hardware architecture and accompanying software (operating system and design tools) are designed to optimize and facilitate the utilization of these platform's fundamental features.

The overall heterogeneous system is built as a set of computation units working as independent processors. They are loosely integrated into the system through identical interfaces and they are connected through a Network-on-Chip (Noc). Each unit represents a subsystem with its own local memory.

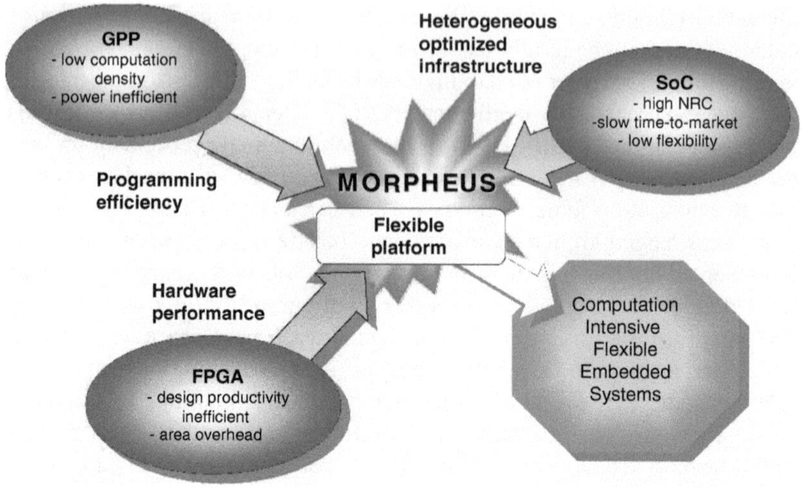

Fig. 1.1 MORPHEUS combines the advantages from FPGA, GPP and SoC approaches

The general purpose processor block which controls the systems represents the user interface toward the system. It is an ARM9 processor core, featuring a standard peripheral set and tightly coupled memories. It runs an embedded Operating System (OS).

The other units connected to the NoC are the reconfigurable engines. They are wrapped to the system as auxiliary processors and loosely coupled thanks to exchange buffers. These exchange buffers are a way to offer a common interface for all these engines in spite of their quite different architectures. This builds the regular system infrastructure which simplifies the design of the platform and its programming. The types of engines selected in the SoC implementation presented here include a 16-bit Coarse Grain Reconfigurable Array (XPP-III from PACT) suitable for arithmetic computation, an embedded FPGA device (the FlexEOS core from M2000) typically designed to handle bit level computation, and a mix-grained 4-bit reconfigurable data-path (the PiCoGA/DREAM reconfigurable processor from ST) suitable for a larger set of applications. They together build a relatively rich heterogeneous architecture offering a large spectrum of capabilities. This architecture is inherently scalable.

A key feature of the approach is to make the final user capable to easily partition its application on the most suitable accelerators to optimize the global performance. This is made possible through a software-oriented approach for the overall synchronization and communication. This includes communication at the system level designed in synergy with the implementation of each function on reconfigurable engine.

Two levels are therefore considered in the proposed programming level. The higher level consists in exploring the application (partitioning in concurrent function kernels, managing their relative dependences, selecting of most suitable fabric, managing their configuration set-up). The lower level consists in exploring the computation implementation on reconfigurable fabrics (implementation design, optimization of data communication with the system). These two levels are quite tightly coupled. The lower level notably provides precise information (parameters, etc.) to the higher level in order to reach a good global efficiency.

Most of the targeted applications have data-stream dominant characteristics with real-time constraints. The platform is made of architecture mechanisms as well as operating system and software tools: it is oriented to permit the optimization of the dataflow running between the hardware units. For example, the platform transparently manages the pipeline of the communications and computations. This is the condition that can allow to sustain the required run-time specification of such type of applications.

Dynamic reconfiguration is an essential aspect of the platform. It corresponds to the high level of the programming model mentioned above. Indeed, the accelerators are not fixed during the utilization of the chip. They are re-initialized each time a new function is required. The dynamic reconfiguration is however done at the reconfigurable engine level. That is to say that the platform does not involve partial reconfiguration within an engine. Two modes are proposed for the allocation of tasks on the accelerators: a mapping at design-time performed by the application programmer or a mapping at run-time performed by the operating system according to the availability of accelerators.

Also, for the case of applications where this data-stream model does not fit so well, the platform allows the programming of parallel threads that can run independently on different accelerators. An Open-MP compliant set of directives can be used to identify some sections of code with such parallel threads.

Moreover, the toolset offers a software-oriented approach for the implementation of an application on the chip. This means that the toolset role is to provide an application design process based on C language as much as possible (also involving a graphical tool for the data-parallelism interconnect aspect of tasks coded with C language). This includes not only the global control of the chip but also the design of the accelerated functions implemented on the reconfigurable engines.

1.3 Intensive Processing Applications Requiring Dynamic Reconfiguration

Applications concerned by this approach are mainly data-stream oriented applications with identified kernels requiring performance acceleration. The presented approach is especially efficient for applications which include (for performance or functional reason) kernels with medium temporal utilization. In the cases where modification rate is high, a more flexible approach based on multi-processor architecture is more efficient. In the other extreme case where the functions are very rarely modified or are even fully static, ASIC solutions can be used. The applications quickly described here are examples that verifies the interest of the approach.

Several applications have indeed been selected in order to specify more precisely the concept, to assess it and to provide quantified measures regarding computing performance, utilization flexibility and implementation productivity.

The emerging IEEE 802.16j standard for mobile broadband wireless access systems is the base for a first type of applications. The standard provides for a baseline PHY chain, with a large number of optional modes, having to do with multiple antenna (MIMO) techniques, or forward error correction (FEC) schemes.

Today's high-end telecommunication networks require data rates up to 40Gbit/s per single line, which cannot be provided by FPGAs or microcontrollers. The solution is the usage of an embedded FPGA (eFPGA) macro that is placed on an ASIC. The design parts, which are considered to be uncertain ("weak parts"), are mapped to the eFPGA, whereas the stable design parts are implemented in ASIC technology.

Cinema productions rely on a huge digital post-processing of the films captured by digital camera or film scanners in resolutions up to 4K (and beyond). The first step in post-processing is film grain noise reduction using motion estimation, temporal Haar filters and discrete wavelet transformations. The algorithm results into up to 2,000 operations per pixel.

Image processing for cameras includes functions such as image enhancement, contour extraction, segmentation into regions, objects recognition, motion detection. Typically, image data rate ranges from 10 to 50 millions of pixels per second and per camera, number of operations ranges from 10^3 to 10^5 operations per pixel so that processing need ranges from 10^{10} to 10^{12} operations per second. In practice it is strongly limited both by technology limits and by price, so that available processing power for an affordable price is a key differentiating factor between competitors. An intelligent camera can be viewed as a large collection of real time algorithms which can be activated in function of non predictable events such as the content of the image or an external information or a request from the user.

1.4 An Heterogeneous Architecture with Dynamically Reconfigurable Engines

The architecture of the MORPHEUS platform is essentially made of

- A control processor (ARM) with peripheral components
- An heterogeneous set of embedded reconfigurable devices of various grain: the three Heterogeneous Reconfigurable Engines (HRE)
- A memory hierarchy and common interface for accelerators
- An efficient and scalable communication and configuration system

All control, synchronization and housekeeping is handled by the ARM9 embedded RISC processor. Computing acceleration is ensured by the three HREs: XPP-III, PiCoGA/DREAM and FlexEOS.

The XPP-III is a coarse grain reconfigurable array primarily targeting algorithms with huge computational demands but mostly deterministic control and dataflow.

Further enhancements based on multiple, instruction set programmable, VLIW controlled cores featuring multiple asynchronously clustered ALUs also allow efficient inherently sequential bit-stream processing.

The PiCoGA/DREAM core is a medium-grained reconfigurable array consisting of 4-bit ALUs. Up to four configurations may be managed concurrently. It mostly targets instruction level parallelism, which can be automatically extracted from a C-subset language called Griffy-C.

The FlexEOS is a lookup-table based, fine grain reconfigurable device. It is a kind of embedded Field Programmable Gate Array (eFPGA). It can map arbitrary logic up to a certain degree of complexity. The FlexEOS may be scaled over a wide range of parameters. The internals of a reconfigurable logic block may be modified to a certain degree according to the requirements.

A homogeneous communication and synchronization mean between each HRE and the rest of the system is provided thanks to the utilization of local dual port/dual clock memory buffers named DEBs (Data Exchange Buffers), CEBs (Configuration Exchange Buffers), and XRs (Exchange Registers). DEBs are utilized for data transfers between system and HREs, CEBs are used to locally store configuration bit-streams of each HRE, while XRs are used for synchronization and control-oriented communicating between the ARM9 and each HRE.

The interconnect mechanism is organized in three separated and orthogonal communication domains: data interconnect, system synchronization and control, configuration management.

AMBA busses are used for control and configurations on one hand and a NoC is used for intense data exchange on the other hand. For efficient data communication, each computation node in the network is provided with an embedded DMA-like data transfer engine that accesses local DEBs and generates relative traffic on the NoC. The same DMA engines are coupled to the storage nodes. In this way, a uniform access pattern is common to each node in the system.

In order to preserve data dependencies in the data flow without having to constrain too much the size and nature of each application kernel, the computation flow can obey to two different synchronization schemes. First scheme is based on explicit synchronization where each computation node is triggered by a specific set of events. In second scheme, the synchronization is implicit, by means of FIFO buffers that decouple the different stages of computation/data transfer. Generally speaking, the XPP-III array appears suited to an implicit synchronization flow, as its inputs are organized with a streaming protocol. The PiCoGA/DREAM accelerator is designed to follow some explicit synchronization. The FlexEOS unit, because of the choices made at the level of the design toolset (see tool chain chapter) also obeys to implicit synchronization. With this system, efficient pipeline on accelerated computations and communications between accelerators can be set-up.

This architecture, thanks to NoC, DEB, DMA-like engines, ARM control, thus facilitates the management of this heterogeneous set of reconfigurable engines. The toolset will complete the concept to make it all the more efficient and to optimize the programming productivity.

1.5 A System-Level Software-Like Tool Chain Including Hardware Targets

For application implementation productivity reason, a target specification of the platform is to put the user in a position of controlling the whole system by programming the ARM core with C language. The design is then aimed at maintaining a high level of programmability, by means of a close synergy between overlapping layers of software libraries and, where necessary, specific hardware services, in order to hide the heterogeneity of hardware units as well of details about data transfer and synchronization details. The heterogeneous architecture can indeed appear as a quite complex object. The goal of the toolset is to make its programming really easy in spite of this complexity (knowing that many elements of the architecture also contribute to the usage simplification).

In order to reach this goal, the programming model is organized in two levels. The high level corresponds to accelerated function, handling macro-operand (source, destination and temporary data of the function) with granularity corresponding to instructions extension, transferred by ARM and controlled by the assistance of a Real Time Operating System (RTOS) providing specific services to manage dynamic reconfiguration (if preferred the end user may also control it through its main program written in C). Hardware resources are thus triggered and synchronized by software routines running on the ARM, either by manual programming or RTOS. The dynamic control of configurations is thus ensured by both SW (RTOS and specific services) and HW (configuration control mechanism). Macro-operands can be data streams, image frames, network packets or different types of data chunks whose nature and size depends largely on the application.

The toolset thus includes compilation of C code in which the programmer includes directives that will identify these accelerated functions. This will permit the automatic management of accelerations, configurations and communications. The programming model implemented by these directives is based on the MOLEN paradigm. In this paradigm, the whole architecture is considered as a single virtual processor, where reconfigurable accelerators are functional units providing a virtually infinite instruction set.

The lower level concerns the internal operation on the accelerators. These operations handle micro-operands that are the native types used in the description of the extension instruction. These types tend to comply with the native datatypes of the specific HRE entry language that is C for ARM, and Griffy-C for PiCoGA/DREAM, Hardware Description Language (HDL) for FlexEOS, NML and FNCPAE-assembly for XPP-III. Micro-operands will only be handled when programming the extensions. In order to keep the design flow globally as close as possible to a software flow, the toolset provides solutions to handle these operations at high level. The programmer can make use of a graphical interface handling boxes for kernel functions. The behavior of these kernels is coded in C language. The toolset provides means for the synthesis of these kernels towards the selected engines. The toolset offers a "retargetable" approach thanks to an intermediate graph level from which low level synthesis is performed. The code towards various architecture targets can indeed be generated from this intermediate level.

This software-like design of accelerator configurations includes the generation of bit-streams for the reconfigurable units, the parameterization of communications (DMA) and their scheduling (pipelined block communication and transformation). In the case of explicit synchronization mechanism, ARM and DMA both contribute to the synchronization of the communication with the computations in the HREs. The toolset role is therefore to program and parameterize these blocks.

Each HRE computation round is applied to a finite input data chunk, and will create an output data chunk. In order to ensure maximum parallelism, during the HRE computation round N following input chunks N+1, N+2, ... should be loaded filling all available space in the DEB but ensuring not to cover unprocessed chunks. Similarly, previous available output chunks ..., N−2, N−1 should be concurrently downloaded ensuring not to access to chunks not yet processed. This can be implemented through the XRs. The DMA is also involved in this process. The toolset role is here to define and optimize the chunk sizes in a coherent manner, taking into account the DEB size and the synthesis of accelerated functions.

For a given function identified as requiring acceleration on a reconfigurable unit, several implementation designs might be available. The selection of an implementation among a library of them is done within a specific configuration file. Besides this selection through configuration file, the capabilities to dynamically select the implementation – and to allocate them to one of the reconfigurable units at run-time – can be done thanks to specific services offered with the RTOS.

1.6 Conclusions

The strength of the presented dynamic reconfigurable platform approach is to build a coherent system (hardware and software) from state-of-the-art technologies, to adapt and integrate them to reach an efficient platform with the benefits of heterogeneous accelerators that can moreover be dynamically reconfigured. The approach introduced here will be more deeply explained in the rest of the book according to the following organization: Several chapters are dedicated to the explanation of the architecture and its various elements. They are followed by the toolset presentation where each module contributes to the simplification of the programmer work. Targeted application examples are then described, including the concept verification made through their implementation on the chip, using the toolset and the board for demonstration. Since this work is part of a European project, the final part of the book deals with the aspect of dissemination, training and exploitation of the results of this project. The management aspect is also described. It provides a good example of organization for this type of research and development activities.

Chapter 2
State of the Art

SoA of Reconfigurable Computing Architectures and Tools

Alberto Rosti, Fabio Campi, Philippe Bonnot, and Paul Brelet

Abstract This chapter provides an analysis of the state of the art about two basic and complementary aspects that drive the development of MORPHEUS: the architecture of a reconfigurable computing platform and the corresponding application toolchain. The two issues are treated in a general manner rather than comparing every single aspect with the MORPHEUS case. So this chapter provides a complete and consistent introduction of the state of the art about reconfigurable computers that can also be read as a standalone contribution being not dependent on the other parts of the book.

Keywords Reconfigurable computers • FPGAs • fine/coarse grain configurable architectures • spatial-timing design • soft-hardware platforms

2.1 Introduction

The concept of reconfigurable computing dates back to the 1960s, when a paper (see references [1,2]) from G. Estrin and his group at the University of California Los Angeles proposed a computer made of one processor and an array of reconfigurable hardware: the UCLA Fixed-Plus-Variable (F + V) Structure Computer. In that historical architecture the main processor was dedicated to the control of reconfiguration, whereas the reconfigurable hardware was used to perform specific tasks such as image processing or pattern matching. It was a hybrid computer architecture combining the flexibility of software with the performances of a hardware solution. That work was triggered by the need to extend the capabilities of computers to handle computations beyond their actual capabilities of that time. Unfortunately

A. Rosti (✉) and F. Campi
STMicroelectronics, Italy
alberto.rosti@st.com

P. Bonnot and P. Brelet
Thales Research & Technology, France

N. Voros et al. (eds.), *Dynamic System Reconfiguration in Heterogeneous Platforms*, Lecture Notes in Electrical Engineering 40,

digital technology was not ready for such a revolutionary change, moreover the relentless increase of performances from microprocessor based solutions, pushed by the improvement of sheer silicon technology, and the need of new programming models and design paradigms inhibited for a long time the commercial development of reconfigurable solutions. From the eighties/nineties it is possible to observe the so called "reconfigurable computing renaissance" with a variety of reconfigurable architectures proposed and developed in industry and academia.

To implement architectures for the demanding computation problems of today there are basically three competing solutions: high-performance microprocessors, application specific integrated circuits and reconfigurable computers. Today the rapidly increasing number of publications at conferences as well as the interest from the academic and industrial community indicates, that reconfigurable computing is gaining acceptance as an intermediate solution between ASICs and microprocessors. High performance processors are multi purpose but they are not enough efficient for certain applications and they have a high power consumption (up to 200 W). ASICs are specific to an application, efficient and low power. Unfortunately they are not general enough and often they are unfeasible because of their manufacturing costs, only high volume products can afford an ASIC design.

Reconfigurable computing is becoming more and more attractive for applications such as cryptography, streaming video, image processing and floating point operations. Precedent publications [3] report that in some case (elliptic curve point multiplication) reconfigurable computing can lead to improvement of 500 times in speed. Another advantage is low power consumption; in certain applications [4] it is possible to obtain 70% savings in power over corresponding microprocessor implementation; typically profiling the application and moving the critical loops to reconfigurable hardware accelerators.

Reconfigurable computers are proposed as the solution of the von Neumann syndrome: the inefficiency of microprocessor based solutions. An unacceptable overhead in execution is in fact due to the instruction-stream based organization of actual computers, the situation will be even worse resorting to multi-processors [5] where there is a proliferation of the processing units and more overhead due to their communication.

Further advantages of reconfigurable computing include reduction in area due to the reuse of the same hardware for computing different functionalities as well as improvements in time to market and flexibility.

A set of surveys [6–14] considering reconfigurable computing from different perspective are listed in the references.

2.2 Reconfigurable Computing Architectures

Reconfigurable architectures are emerging structures of computers that combine the programmability of processors with spatial design. They are a class of architectures that allows adapting to the instantaneous needs of an application, obtaining specialization of performances, flexibility and power consumption also at run time. In this section we are going to analyze their basic features.

Several approaches for reconfigurable architectures have been proposed. Typically they can be generalized by the idea to extend a main general purpose processor by one or more reconfigurable processing units. Even though such reconfigurable processing units could theoretically run standalone, they are often coupled with a general purpose processor, since the reconfigurable units are not universal or powerful enough to handle complete applications.

2.2.1 *Basic Terminology*

It is necessary at this point to introduce some basic terminology because the complexity and variety of the cases encountered in this analysis could lead to misunderstandings.

2.2.1.1 Configurability

A configurable architecture is an architecture that can be forged into different shapes at design time depending on the value assigned to its configuration parameters.

Configuration can happen at different levels of abstractions: architectural, or micro-architectural. Architectural configuration implies that the actual programming view of the user is changing. For instance an extension of the ISA can occur by introducing special instructions that are executed on the configurable processing elements. Configuration at the micro-architecture level implies that the organization of the functional units is affected.

2.2.1.2 Reconfigurability

A reconfigurable architecture can be customized after fabrication by changing its logical structure to implement a different functionality while its physical structure remains unchanged.

Reconfigurability can be static or dynamic. In the case of static reconfigurability an architecture can be configured several times for different functionalities at load time. In case of dynamic reconfiguration an architecture can switch the application at execution time.

Reconfigurable computing is data-stream-based and inherently parallel, it should be clearly featured as the counterpart of von Neumann architectures which are instruction-stream-based and sequential. Reconfigurable computers (at least statically) have no instruction fetch and program counter, they have data counter instead.

Reconfigurable computers are different than multiprocessors, also run time connect switching of a von Neumann processors is not comparable to reconfiguration.

2.2.1.3 Instruction vs. Stream Based Computing

For their ability to be programmed reconfigurable architectures are compared with general purpose processors. There is a first fundamental difference indeed: in pure software solutions a program is made of a sequence of instructions whereas a reconfigurable computer is programming the structure of the computer itself. A configuration code (configware) is used instead of software, programming is made in space rather than in time. Another fundamental difference is that reconfigurable architectures are inherently parallel, so they are well suited to work independently on data streams (flowware). This classification scheme is taken from http://en.wikipedia.org/wiki/Reconfigurable_computing by N. Tredennick.

2.2.1.4 Time and Spatial Computation

Reconfigurable computing is characterized by finding a trade off between time and spatial computation. Spatial implementations are typical of hardware; a computation resource exists at all the points in space where it is needed, allowing maximum parallelism of computation and a large but efficient solution. In temporal computation, typical of software artifacts, a small amount of resources is reused in time when needed, leading to a more compact but less efficient solution. With reconfigurable solutions it is possible to tune a mixed spatial–temporal computation approach, exploiting more efficiency related to hardware.

2.2.1.5 Binding Time

A computing system can be generally conceived as a set of operations mapped onto computation resources: this process of mapping is called binding. Binding time is a fundamental feature that can be used to analyze computers comparing reconfigurable computers to their counterparts: ASICs and microprocessor based solutions. In hardware solutions every operation is associated directly to a computation resource, binding is performed statically at fabrication time with no overhead. In the case of software implemented on a microprocessor, operations are described (coded) by instructions that manipulate data. Binding is achieved by the instruction decoding and execution mechanism that solves the complex mapping between an instruction and its needed computation resources. Mapping from an instruction to the computation resources is resolved dynamically at every clock cycle, requiring a large overhead in terms of performance and power. With reconfigurable solutions a new flexibility is added to binding because it is performed at loading time. Large overhead due to stream-of-instruction organization are completely avoided.

2.2.2 Reconfigurable Computing Devices

In this section we are going to analyze the different kinds of reconfigurable devices that are in use today.

2.2.2.1 FPGAs

FPGAs were the first devices which commercially introduced a new class of computing architectures based on reconfigurability. These architectures can be customized to specific problems after fabrication. As hardware solutions they exploit spatial computation to increase performances.

FPGA are fine grained reconfigurable devices which provide great flexibility and limited improvement in computing performances.

In fine grained reconfigurable fabric, functional units implement a single bit or a few bits function based on simple lookup tables that are usually organized in clusters and are interconnected. Fine grained architectures are very flexible but compared to coarse grained architectures consume more area due to fine interconnect and are less efficient for computation because of the area overhead and poor routability; they can be efficiently used for applications with high weight on control flow. For this kind of reconfigurable computers design flows are similar to a hardware design flow involving development in HDL generating configuration code.

2.2.2.2 Coarse Grained Devices

Reconfigurable computers based on coarse grain logic processing elements are oriented to specific application, they obtain an increased performances compared to FPGAs but at the expense of flexibility. They provide high potential to speed up data streaming applications characterized by high data parallelism.

Efficiency of coarse grained reconfigurable hardware comes from the regular structure of their configurable functional blocks (usually ALUs and memory elements), they have also a simpler data path and routing switches working at word level.

Mapping of logical functionality onto a coarse grained reconfigurable architecture is also simpler that on a fine grained architecture. Operations can be more naturally match to the capabilities of the processing elements.

The final benefit of coarse grain vs. fine grained architectures is the reduced size of the configuration memory and the reduced complexity of place and routing.

For this kind of reconfigurable computers design flows are similar to a microprocessor programming flow with compilation generating assembly language.

2.2.2.3 Special Purpose Configurable Platforms

Configurable or reconfigurable computing is also used for a few special purpose computing architectures.

Supercomputers

Reconfigurable solutions are currently used for commercial products by high performance computers (supercomputers). Those systems, such as Cray XD1 and

SGI RASC generally use multiple CPUs in parallel and augment the performances for application specific computing adding large FPGAs.

Hardware Emulators

Emulation systems are historically important as an example of use of reconfigurability to emulate the hardware models. Quickturn was an important example in the nineties, more recent implementations are Palladium from Cadence design systems and Vstation Pro station from Mentor Graphics.

Configurable Instruction Set Processors

Another approach to meet the special requirements of the different application domains are configurable instruction set processors as offered by ARC and Tensilica's Xtensa technology. They are von Neumann architecture with some configuration capability added on in the ALU for changing or modifying the instruction set. These configurable processors need to be configured at design time.

2.2.3 *Examples of Reconfigurable Computing Platforms*

This section contains a list of exemplar reconfigurable architectures from the classes just defined in Section 2.2.2.

Xilinx Virtex®-5 FPGAs are the world's first 65-nm FPGA family fabricated in 1.0v, triple-oxide process technology, providing up to 330K logic cells, 1,200 I/O pins, 48 low power transceivers, and built-in PowerPC® 440, PCIe® endpoint and Ethernet MAC blocks, depending upon the device selected.

Altera Stratix® IV 40-nm FPGAs is a high-end FPGA delivering high density up to 680K logic elements, 22.4Mbits of embedded memory, and 1,360 18 × 18 multipliers, high performance and low power.

The GARP chip [15] is designed by the BRASS (Berkley Reconfigurable Architectures, System & Software) research group. It combines a standard MIPSII processor with a dynamically reconfigurable array of (32 rows by 24 columns) of simple computing elements interconnected by a network of wires implemented on FPGAs. Every element includes 4 bits of data state is a 2-bit logic blocks that takes four 2-bit inputs producing up to two 2-bit outputs. The 24th column of control blocks is dedicated to manage communication outside the array, the architecture proposes a direct connection between the reconfigurable array and the memory.

DP-FPGA [16] is a proposed architecture where an attempt has been made to mix fine grain and coarse grain data-paths in order to implement reconfigurable structured data paths.

RAW (Reconfigurable Architecture Workstation) [17] provides a regular multi-core signal processing scalable platform. Raw is a tiled multicore architecture made of 16 32-bit modified MIPS R2000 microprocessor in a 4 × 4 array. Each tile is comprising instruction, switch-instruction, and data memory, an ALU, FPU, registers, a dynamic router, and a programmable switch. The tiles are interconnected by an on chip network, each tile is connected to its four neighbors. The RAW architecture is aimed at exploiting different kinds of parallelism at instruction, data task and streaming level.

Imagine [18] is a programmable single-chip processor that supports streaming programming model for graphic and image processing applications. This architecture is made of 48 ALUs organized as 8 SIMD clusters of 6 ALUs, executing static VLIW instructions.

MorphoSys [19] is a coarse grain, integrated and reconfigurable system on chip for high throughput and data parallel applications. It is made of a reconfigurable array of processing cells, a control processor and data/configuration memories. It has been applied for video compression and data encryption applications.

MS-1 is a coarse grain architecture from Morpho Technologies (delivered to Freescale as MRC6011) implementing 16-bit to 32-bit processing elements. The routing architecture follows the MorphoSys-like approach in which the processing elements can communicate only with nearest-neighbour cells and a regional (hierarchical) connections.

DAPDNA-2 is a coarse grain architecture from IPFlex that implements 16-bit to 32-bit bit processing elements, it provides fast context switch among different configurations. It features 6 32-bit wise I/O channels.

D-Fabric from Elixent's technology (recently acquired by Matsushita Electronics) is an example utilizing a coarse architecture. It features a dedicated 4-bit ALU implementing sum/subtraction providing high flexibility in terms of resource utilization especially for non-standard operand sizes.

RaPiD (Reconfigurable Pipelined Datapath) [20] is a coarse-grained reconfigurable architecture specialized for signal and image processing. It provides a reconfigurable pipelined datapath controlled by efficient reconfigurable control logic.

PipeRench [21] is a reconfigurable fabric made of an interconnection of configurable logic and storage elements that can be combined to a DSP, microcontroller of general purpose processor for accelerating streaming media applications. It provides fast context switch among reconfiguration.

The Pleiades Architecture [22] is a crossbar based structure built on an architecture template where a control processor drives an array of autonomous reconfigurable satellite processors that communicate over a reconfigurable network. The communication is managed by a data driven computation model. The data intensive parts of DSP applications are executed on the satellite processors.

The **ADRES** architecture [23] by IMEC combines the capabilities of a VLIW along with the benefits of a reconfigurable array. It maps at thread level on one of the two computation resources.

Montium is a tiled reconfigurable microprocessor architecture adapting itself to the actual use and environment; it is designed for low power mobile applications.

Recore Systems is a fabless semiconductor company that develops advanced digital signal processing platform chips and licenses reconfigurable semiconductor IP.

HARTES holistic approach to embedded applications requiring heterogeneous multiprocessor platforms RISC, DSP, including reconfiguration. It is aimed at bridging the gap between SW and architecture, focusing on methodology and tool-chain.

4S Smart Chips for Smart Surroundings, develops flexible heterogeneous (mixed Signal) platform including analogue tiles, hard-wired tiles, fine and coarse-grained reconfigurable tiles, Microprocessors and DSPs linked through a Network on Chip.

2.3 Methods and Tools for Embedded Systems

This section highlights the evolution of state-of-the-art in the domain of methods and tools for reconfigurable computing. It also permits to check and confirm the relevance of the proposed solutions regarding this state-of-the-art. It mainly focuses on showing the current trends, and does not present advanced research subjects but established tools and methods.

The MORPHEUS project focuses on tools allowing the development of applications on reconfigurable architectures. In the MORPHEUS tools context, the system architecture is pre-defined.

It reminds here that the proposed toolset permits to program the MORPHEUS circuit thanks to a two-level approach. The highest level is a C language program in which some functions (identified before entering the toolset) can be replaced by accelerated operations implemented on reconfigurable units. This first level permits to address the high level control-dominant part of the application. It is also at that level that a RTOS is proposed to manage dynamic reconfiguration.

The second level is the description of these functions. Since they are implemented on reconfigurable units, they are supposed to contain some intrinsic parallelism. That is why, for example, a graphic solution allowing the expression of this parallelism is considered and proposed.

A current trend is the increase of the abstraction level of EDA tools. So, this section starts by presenting proposed model and language. Then, it gives an overview of tools managing design at system level. We also here present state-of-the art on both EDA and compilation that could be interesting for reconfigurable computing. Finally, according to the presented outline, the concluding section of this chapter shows that the proposed MORPHEUS solution appears to stay valid, relevant and competitive.

2.3.1 High-Level Languages and Models Expressing Parallelism

The objective of this section is to provide an overview of languages supporting computation models enabling an efficient expression of time and/or space parallelism.

The grail conquest here consists in searching a way to conciliate on one hand the efficient expression of the parallelism that can be inherent within an application (or the idea that engineers can have of a parallel implementation of it), with, on the other hand, the programming easiness that is often considered as requiring an easy and quick understanding by software engineers used to languages like C++ or Java.

2.3.1.1 System & Application-Oriented C-Based Modeling

Some extensions of C/C++ deal with System level modeling.

Ratified as IEEE Std. 1666™-2005, **SystemC™** is a language built in standard C++ by extending the language with the use of class libraries. A few commercial tools such as **Coware ConvergenSC**, **Synopsys System Studio, ARM, Cadence, Intel, Mentor Graphic** are based on SystemC.

SystemVerilog IEEE 1800™ likewise is a rather new language built on the widely successful Verilog language and has emerged as the next-generation hardware design and verification language.

Many works have addressed the problem of HW synthesis from C language, by language extension or by adding annotation in the code.

Mitrion-C is a parallel C-family programming language to fully exploit the parallel nature of the Mitrion Virtual Processor.

SA-C (by Colorado State Univ. within Cameron project) is a variant of the C programming language that exploits instruction-level and loop-level parallelism, arbitrary bit-precision data types, and multidimensional arrays.

Impulse C (Impulse Accelerated Technologies) is a subset of the C programming language combined with a C-compatible function library supporting parallel programming, in particular for programming of applications targeting FPGA devices.

2.3.1.2 Stream-Oriented Languages and Models & Other References

Stream oriented languages are dedicated to modeling digital signal processing applications, and are more specific languages.

MATLAB/SIMULINK (see later subsection on EDA tools) allows to model applications as a data flow diagram, with a set of predefined block.

Scilab is a scientific software package for numerical computations providing a powerful open computing environment for engineering and scientific applications.

Array-OL (Thales Underwater Systems/LIFL) is a language to describe parallelism in data stream signal processing applications.

The reader could also refer to Stream-C, ASC, SNET languages.

Beyond the above described languages, it is interesting to quote a few API devoted to the expression of parallelism like **DRI** (Data Reorganization Interface) or **VSIPL/VSIPL++, MPI, OpenMP, OpenUH, Stream Virtual Machine API** in PCA DARPA program. **UML, SysML** address the problem of formalizing specification requirements.

Many different languages or dialects for parallelism description and hardware-oriented implementation of computation kernels have been proposed. Almost all of these languages are based on C, essentially for reasons related to user friendliness and legacy. While some sort of simplified C-based language with pragma or annotations has proved efficient for describing parallelism in computation, often data transfer and organization has appeared as difficult to express with standard C-based notations. For these reasons, more specific languages oriented on data transfer virtualization in form of streams have also been proposed to target signal processing applications.

2.3.2 System Level Design Tools

At system level, different kinds of tools are available for algorithm development, parallelization and mapping of applications on a target architecture, HW/SW co-design or formal verification. Some tools have also addressed at the system level the specificities of reconfigurable systems.

2.3.2.1 Algorithmic Tools & Application Parallelization and Mapping

The **PTOLEMY II** toolset (by Berkeley) is interesting because of the numerous models it proposes and the possibility to make them interoperate. The Ptolemy project studies heterogeneous modeling, simulation, and design of concurrent systems.

Matlab/Simulink (by TheMathworks) is a toolset for signal processing algorithm development and tuning.

R-Stream compiler (by Reservoir Labs) is a "High Level Compiler" targeting embedded signal/knowledge processing with high performance needs, and is designed to seamlessly plug into target-specific (low-level) C compilers.

Gaspard (by LIFL) is a development environment for a high level modeling of applications in Array-OL, applications transformations and mapping on an architecture.

2.3.2.2 Tools Oriented to Reconfigurable Computing & Formal Specification

In this subsection, tools aiming specifically at reconfiguration aspects will be considered. This can be for example dynamic reconfiguration management or others aspects like management of reconfigurable units within a computing system. This category of tools is only present in the academic domain.

The **EPICURE** project proposes a design methodology able to bridge the gap between an abstract specification and a heterogeneous reconfigurable architecture.

The **RECONF** IST project proposed a high-level temporal partitioning algorithm, which is able to split the VHDL description of a digital system into two equivalent sub-descriptions.

SpecEdit (Chemnitz/Lucent) is an extensible framework to support the creation of specifications for technical systems. So, this tool is complementary of a model checker tool like Cadence **SMV.**

2.3.2.3 Compilers

Traditional compilation tools and operating systems will be used by MORPHEUS to exploit the ARM-based processor environment as system control and synchronization engine with the highest degree of user-friendliness and efficiency. Of course, SW methods and tools are also impacted by the reconfigurability of a system, as the reconfiguration mechanism and accelerated functions mapped on reconfigurable units must be managed by software calls, via operating system, middleware, and specific hardware services such as interrupts, DMA or configuration managements.

The **Streams-C** compiler synthesizes hardware circuits for reconfigurable FPGA-based computers from parallel C programs.

Mescal Architecture Description Language (MADL) (by Princeton Univ. within Mescal project) is used for processors modeling in order to provide processor information to tools such as compilers and ISS.

2.3.3 C-Oriented Synthesis Tools

Synthesis tools are not literally tools utilized for application development but we have classified them into this category because, in the MORPHEUS context, they need to feature a close integration with source code compilation and software development. In particular, in the MORPHEUS toolset, hardware synthesis on the computation units starting from high level descriptions can be considered the natural extension of compilation tools towards the implementation of kernels over the reconfigurable units.

Regarding High level synthesis methodologies, two different solutions can be distinguished: design exploration tools and compilation/synthesis tools. Whereas design implementation at low/RTL level is based on logic synthesis, system level solutions are currently mainly based on interactive design space exploration.

2.3.3.1 Design Space Exploration and ASIP Synthesis

LisaTek (by Coware) uses the LISA language to generate both a specific processor (RTL code) and its associated SW development environment (assembler, linker, archiver, C compiler and ISS).

Chess/Checkers (by Target Compiler) is also a retargetable compiler that starts from the C code of the application to design a processor completed by a set of acceleration extension.

XPRES Compiler (by Tensilica) is the development environment for their Xtensa processor optimization.

2.3.3.2 Hardware Mapping from High-Level Languages

Numerous products can be found in this category: technology independent solutions are generally based on a hardware oriented design approach (this is the case of C-based synthesis of circuits for ASIC/FPGA/eFPGA) to which it can be possibly linked a more system-oriented approach. Such tools make use of synthesis-oriented steps, which require fine grained computation fabric as a target for making efficient use of the underlying technology. On the contrary, mapping tools for coarse-grained fabrics tend to be target-oriented, and generally propose device specific dedicated approaches, very often based on standard compilation techniques rather than logic synthesis.

CatapultC (by Mentor Graphics) permits to synthesize VHDL code from a C description. The C code has to be designed with care for this purpose.

AccelDSP (by Xilinx) synthesizes DSP blocks for Xilinx FPGA from a MATLAB description.

FELIX (by University of Karlsruhe) is a design space exploration tool and graphical integrated development environment for the programming of coarse-grained reconfigurable architectures.

GARP [15] (Berkeley BRASS group). Applications are written in C.

The **ADRES** [23] architecture (by IMEC) makes benefit of its DRESC toolset. Threads are time multiplexed on one or the other type of model with the help of the proposed tool.

The **FPOA** (Field Programming Object Array by MATHSTAR) comes with a toolset combining SystemC and a Verilog-like language named OHDL.

2.4 Conclusions

MORPHEUS aims to provide **best of class solutions for embedded systems.** Existing commercial solutions are so far mostly focused on fine-grained FPGA devices, that bring limited benefits in combining flexibility (field programmability) and efficiency (computing density, development time). Alternative solutions, based on coarse-grained logics are application-oriented and often lack generality, proving unable to adapt themselves to different bit-width and different computation pattern. MORPHEUS provides an integrated solution that could **enhance and combine the most promising approaches to post-fabrication customizable hardware in an integrated platform concept.** The heterogeneous set of programmable computation units adapt to a broad class of applications. The MORPHEUS architecture is composed by the following main entities: an ARM9 processor, bus architecture, peripherals and IO system that act as chip control, debug, synchronization and main interface, an interconnect architecture based on a NoC, a set of HREs that act as computation resources for the chip.

As a complex, multi-core heterogeneous SoC for DSP-oriented computation MORPHEUS can hardly be compared with the rich landscape of state-of-the art reconfigurable fabrics. But as a single chip entity, MORPHEUS should be more properly compared to large "general purpose" multi-core platforms (homogeneous and heterogeneous) for intensive digital signal processing, such as commercial FPGA-based systems like Xilinx Virtex II or Altera Stratix on one side, DSP oriented computational engines such as OMAP by TI, STW51000 by ST or Intel PXA, and big, interconnect-centered multicore platforms such as the RAW [17] architecture, Pleiades [22], the PicoChip, IP-Flex Imagine [18] processor or the Cell [24] processor. In comparison with FPGA-based systems like Stratix or Virtex, MORPHEUS can offer a larger computation density thanks to the contribution of medium/coarse-grained programmable units.

With respect to regular multi-core signal processing platforms such as RAW [17], Cell [24], Imagine [18] or PicoChip, MORPHEUS presents an architecture that is strongly heterogeneous. It is definitely less scalable, but it compensates for that as it offers computing units that feature coarse computation density and large processing power. In turn, while the number of computation units and their organization is not easily scalable, the size of computation units themselves and the interconnect mechanism can be parametric, thus offering a degree of customization that can be considered analogous to the scalability of the number of units in more regular architectures. On the other hand, the reduced numbers of units eases data communication and interconnect strategy, making it possible for the controlling processor to synchronize explicitly every required computation or data transfer.

MORPHEUS contributes to the state of the arts also as far as tools and design methodologies are concerned. Parallel languages that can be currently identified do not appear to bring a clear advantage in term of standardization, parallelization efficiency and programming easiness. The MORPHEUS approach is mainly based on a C language description for the system level description of the application implemented on the platform, a graphical description of the data-stream dominant functions accelerated on the reconfigurable units including sub-function whose behavior is described in C language too. The top border of MORPHEUS tools has been defined as the output of "high level compilers" whose role would be to produce the mapping of an application description towards a roughly defined architecture, the development of these tools is out of the scope of the project.

Tools addressing generic issues of reconfigurable computing (global system design including dynamic reconfigurability and space/time parallelization) are not numerous and the MORPHEUS solution addresses well all these aspects. The global aspect system is seen in MORPHEUS through the MOLEN paradigm efficiently completed by OS and hardware configuration management for dynamic reconfiguration.

Existing high-level synthesis tools are either limited to fine-grain architectures, or are target specific in case of coarse grain architectures. The spatial design tools of MORPHEUS are not specific to a target, and allow abstracting the architecture heterogeneity.

Technology specific design issues are handled, according to the MOLEN paradigm, and through the Madeo retargetable synthesis by product-specific mapping tools (GriffyC for PiCoGA/DREAM, XPP compiler for Pact XPP, FlexEOS place and route engine for M2000).Technologies involved in MORPHEUS are at a state-of-the-art level in each of their domain. This is what can be observed for instance for MOLEN, CoSy, CASCADE, MADEO, SPEAR, SpecEdit. Doing a global and comprehensive flow is a challenge and requires filling the existing gap between tools. A significant element of innovation related to MORPHEUS is the integration of these existing approaches in one design flow.

The main benefit of MORPHEUS to the reconfigurable computing domain is that it introduces the ability to address new classes of embedded applications with high dynamic behavior, allowing dynamical shaping of the application on a heterogeneous reconfigurable architecture. MORPHEUS is close to providing a general purpose platform by a clever mix of granularities and tools – going beyond platform FPGAs which are domain-specific.

References

1. G. Estrin, Reconfigurable computer origins: the UCLA fixed-plus-variable (F + V) structure computer, *IEEE Annals of the History of Computing* 24, 4 (Oct. 2002), 3–9.
2. G. Estrin, Organization of Computer Systems—The Fixed Plus Variable Structure Computer, *Proceedings of Western Joint Computer Conference, Western Joint Computer Conference*, New York, 1960, pp. 33–40.
3. N. Telle, C.C. Cheung and W. Luk, Customising hardware designs for elliptic curve cryptography, *Lecture Notes in Computer Science*, 2004, 3133.
4. G. Stitt, F. Vahid and S. Nematbakhsh, Energy savings and speedups from partitioning critical software loops to hardware in embedded systems, *ACM Transaction on Embedded Computer Systems*, 2004, 3(1), pp. 218–232.
5. S.K. Moore, Multicore is bad news for supercomputers, *IEEE Spectrum*, November 2008.
6. T.J. Todman, G.A. Constantinides, S.J.E. Wilton, O. Mencer, W. Luk and P.Y.K. Cheung, Reconfigurable computing: architectures and design methods, *IEE Proceedings-Computers and Digital Techniques*, 152(2), March 2005, 193–207.
7. P. Schaumont, I. Verbauwhede, K. Keutzer and M. Sarrafzadeh, A quick safari through the reconfiguration jungle, *Proceedings of the 38th Design Automation Conference (DAC) 2001*, Las Vegas, NV, USA, June 18–22, 2001, pp. 172–177.
8. R. Hartenstein, A decade of reconfigurable computing: a visionary retrospective, *Proceedings of DATE '01*, Munchen, March 13–16, 2001.
9. A. DeHon and J. Wawrzynek, Reconfigurable computing: what, why, and implications for design automation, *Proceedings of the 36th Design Automation Conference (DAC) 1999*, New Orleans, Louisiana, USA, June 21–25, 1999.
10. K. Bondalapati and V.K. Prasanna, Reconfigurable computing systems, *Proceedings of IEEE*, 2002, 90(7), pp. 1201–1217.
11. K. Compton and S. Hauck, Reconfigurable computing: a survey of systems and software, *ACM Computing Surveys*, 2002, 34(2), pp. 171–210.

12. W. Luk, P.Y.K. Cheung and N. Shirazi, Configurable computing, *in Chen, W.K. (Ed.): Electrical Engineer's Handbook*, Academic Press, 2004.
13. R. Tessier and W. Burleson, Reconfigurable computing and digital signal processing: a survey, *Journal of VLSI Signal Processing*, 2001, 28, pp. 7–27.
14. C. Bobda, *Introduction to Reconfigurable Computing: Architectures, Algorithms and Applications*, Springer Verlag, 2007.
15. J.R. Hauser and J. Wawrzynek, Garp: a MIPS processor with a configurable coprocessor, *Proceedings of FPGAs for Custom Computing Machines*, NAPA Valley, CA, USA Apr. 16–18, 1997, pp. 12–21.
16. D. Cherepacha and D. Lewis: a datapath oriented architecture for FPGAs, *Proceedings of FPGA '94*, Monterey, CA, USA, February 1994.
17. M. Bedford Taylor, W. Lee, J. Miller, D. Wentzlaff, I. Bratt, B. Greenwald, H. Hoffmann, P. Johnson, J. Kim, J. Psota, A. Saraf, N. Shnidman, V. Strumpen, M. Frank, S. Amarasinghe, and A. Agarwal, Evaluation of the raw microprocessor: an exposed-wire delay architecture for ILP and streams, *Proceedings of 31st Annual Intenational Symposium on Computer Architectures (ISCA)* 2004, pp. 2–13.
18. J.H. Ahn, W.J. Dally, B. Khailany, U.J. Kapasi and Abhishek Das, Evaluating the imagine stream architecture, *Proceedings of the 31st Annual International Symposium on Computer Architecture*, Munich, Germany, June 2004.
19. H. Singh, M.-H. Lee, G. Lu, F.J. Kurdahi, N. Bagherzadeh and E.M. Chaves Filho, MorphoSys: an integrated reconfigurable system for data-parallel and computation-intensive applications, *IEEE Transactions on Computers*, May 2000.
20. C. Ebeling, C. Fisher, G. Xing, M. Shen and Hui Liu, Implementing an OFDM receiver on the RaPiD reconfigurable architecture, *IEEE Transactions on Computers*, 53(11), November 2004, pp. 1436–1448.
21. S.C. Goldstein, H. Schmit, M. Budiu, S. Cadambi, M. Moe and R. Reed Taylor, PipeRench: a reconfigurable architecture and compiler, *IEEE Computer*, April 2000.
22. H. Zhang, et al., A 1V heterogeneous reconfigurable processor IC for baseband wireless applications, *ISSCC Digest of Technical Papers*, 2000, pp. 68–69.
23. B. Mei, S. Vernalde, D. Verkest, H. DeMan and R. Lauwereins, ADRES: an architecture with tightly coupled VLIW processor and corse-grained reconfigurable matrix, *Proceedings of FPL 2003*.
24. J.A. Kahle et al., Introduction to the Cell Multiprocessor, *IBM Journal of Research and Development*, 49(4/5), July 2005, pp. 589–604.

Part II
The MORPHEUS Architecture

Chapter 3
MORPHEUS Architecture Overview

Wolfram Putzke-Röming

Abstract This chapter provides an overview of the MOPRHEUS platform architecture. Moreover, it discusses and motivates several of the architectural decisions made during the development of the platform.

Keywords ARM processor · configuration manager · prototype chip · DREAM · platform architecture · M2000 · load balancing · NoC · XPP

3.1 Introduction

Starting point for the MORPHEUS platform architecture was the idea to define a reference platform for dynamic reconfigurable computing to be efficiently used in different application domains. In particular, real-time processing is in the focus of this platform. It is obvious that flexibility, modularity, and scalability of such a platform are key requirements in order to allow an efficient adaptation of the platform architecture to the specific requirements of a certain application.

Another fundamental and new idea was to integrate heterogeneous, reconfigurable computation engines (HREs), which support different but complementary styles of reconfigurable computing, in one platform. Three state-of-the-art dynamically reconfigurable computation engines, representing fine-grain, mid-grain, and coarse-grain reconfigurable computation architectures, have been selected and integrated into the MORPHEUS platform.

In summary, the goal of the MORPHEUS platform architecture is to combine the benefits of the different styles of reconfigurable computing in one platform. Since the platform is designed to be highly flexible and scalable, different applications from various application domains can be addressed with the platform.

W. Putzke-Röming (✉)
Deutsche Thomson OHG, Germany
wolfram.putzke-roeming@thomson.net

N. Voros et al. (eds.), *Dynamic System Reconfiguration in Heterogeneous Platforms*,
Lecture Notes in Electrical Engineering 40,

3.2 Prerequisites and Requirements

Two fundamental design decisions regarding the MORPHEUS platform architecture were made at an early stage of the project. The first decision was to use a central controller to manage the whole platform. The second decision was the usage of the Molen paradigm [2] to control the HREs from the central processor.[1] From an architectural perspective the Molen paradigm uses registers, which are called exchange registers (XRs), to control the processing of the reconfigurable engines and to pass parameters.

A first analysis of the selected test applications showed quickly that in general two different principles of data processing have to be supported. The first principle naturally targets data stream processing. If the application allows that consecutive processing steps can be mapped onto different HREs an execution pipeline across multiple HREs can be created. (Stage1: load data into HRE1, stage2: process data by HRE1, stage3: transmit results from HRE1 to HRE2, etc.) Thereby, the difficult task is to find a well balanced split of the application to the pipeline stages, since the execution speed of the whole pipeline is limited by the slowest stage. The second principle is based on the repeated usage of the same HRE for consecutive processing steps. This principle requires that sufficiently large memory is available to store intermediate results between the consecutive runs of the HRE. Moreover, a certain reconfiguration overhead has to be accepted, as the HRE must be reconfigured for every processing step. Since this approach uses only one processing engine the resource utilization of the other available processing engines is not optimal. Of course, also a mixture of both processing principles is possible and can be used to find a good load balance for all available HREs. In Chapter 8 the targeted models of computation for the MORPHEUS platform are discussed in more detail.

Furthermore, the analysis of the applications clearly showed that special emphasis has to be put to the dynamic reconfiguration mechanism. Especially for the second processing principle it can be necessary to change the configurations of the three reconfigurable processing engines frequently. Thus, an appropriate solution is required to minimize the reconfiguration time and related processing load for the central control processor. Depending on the application and its mapping onto the MORPHEUS platform the requirements regarding the maximum reconfiguration time can vary significantly. However, especially if real-time processing is required, the reconfiguration time must be in the area of a few microseconds or even less.

---

[1] The central control concept of the MORPHEUS platform was chosen in order to simplify the hardware and software design. However, it has to be emphasized that this decision does not limit the very flexible usage of the HREs (e.g. dynamic load balancing controlled by an operating system).

3.3 The MORPHEUS Platform Architecture

The MORPHEUS hardware architecture comprises three heterogeneous, reconfigurable processing engines (HREs) which target different types of computation. All HREs are presented and discussed in more detail in Chapters 4–6:

- The PACT XPP is a coarse-grain reconfigurable array primarily targeting algorithms with huge computational requirements but mostly deterministic control- and dataflow. Recent enhancements also allow efficient sequential bitstream and general purpose-processing. These enhancements are based on multiple, instruction set programmable, VLIW controlled cores which are equipped with multiple asynchronously clustered ALUs.
- DREAM is based on the PiCoGA core from ARCES. The PiCoGA is a medium-grained reconfigurable array consisting of 4-bit ALUs and 4-bit LUTs. The architecture is mostly targeting instruction level parallelism, which can be automatically extracted from a C-subset language called Griffy-C. The DREAM mainly targets computation intensive algorithms that can run iteratively using only limited local memory resources.
- The M2000 is an embedded Field Programmable Gate Array (eFPGA). Thus, it is a fine-grain reconfigurable device based on LUTs. It is capable to be configured with arbitrary logic up to a certain level of complexity.

All control, synchronization, and housekeeping is handled by an ARM 926EJ-S embedded RISC processor. It shall be emphasized that the prime task of the ARM processor is to be the central controller for the whole platform.[2]

As the HREs in general will operate on differing clock domains, they are decoupled from the system and interconnect clock domain by data exchange buffers (DEB) consisting of dual ported (dual clock) memories either configured as FIFOs or ping-pong buffers.[3] From a conceptual point of view the HREs can access their input data only from their respective local DEBs. The ARM processor, which is in charge of controlling all data transfers between memories and DEBs or between DEBs, has to ensure the in-time delivery of new data to the DEBs to avoid idle times of the HREs.

According to the Molen paradigm each HRE contains a set of XRs. Through the XRs the ARM and HREs can exchange synchronization triggers (e.g. new data has been written to DEBs or the computation of HRE has finished) as well as a limited number of parameters for computation (e.g. start address of new data in the DEBs or parameters that are necessary for the interpretation of the data).

---

[2] Other computation tasks should be mapped onto the ARM only if its role as central system controller is not affected.

[3] Ping-pong buffering is a mechanism to avoid idle times of the HREs while they are waiting for new data. Ping-pong buffering requires an even number of input/output buffers. If only one buffer is available it is necessary that this buffer allows parallel read and write accesses. While the HRE processes the data of the "ping" buffer, new data is pre-loaded into the "pong" buffer.

The buffering of local data can be done in the on-chip data memory. This SRAM may either be used as a cache or scratchpad RAM. To satisfy the high application requirements regarding memory throughput an advanced DDRAM controller provides access to external DDR-SDRAM. A recommendation for an appropriate DDR-SDRAM controller is given in [3].

To summarize, the MORPHEUS platform architecture has a three level memory subsystem for application data. The first level, which is closest to the HREs, is represented by the DEBs. The second level, which is still on-chip, is the on-chip data memory. Finally, the third level is the external memory. In Chapter 8 the three level memory subsystem is presented with more detail.

As dynamic reconfiguration of the HREs imposes a significant performance requirement for the ARM processor, a dedicated reconfiguration control unit (PCM) has been designed to serve as a respective offload-engine. The PCM analyzes which configuration is needed for the next processing steps on the HREs. Depending on this analysis the next configurations are pre-loaded. It should be mentioned that the memory subsystem used for handling configurations also uses the same three layer approach previously introduced for the application data. The configuration exchange buffers (CEB) inside of the HREs are the first layer. The second layer is the on-chip configuration memory, and the third layer is the external memory.

All system modules are interconnected via multilayer AMBA busses. Separate busses are provided for reconfiguration and/or control and data access. As the required bandwidth for high-performance and data intensive processing might become quite high, an additional network on chip (NoC) based on ST's spidergon technology [1] has been integrated. To reduce the burden on the ARM system controller, DMAs are available for loading data and configurations. However, data transfers on the NoC also have to be programmed and initiated by the ARM processor. Similar to programming a DMA for the AMBA bus, the ARM can program the DNA (direct network access) module for NoC transfers.

Figure 3.1 does not provide details regarding the NoC. It simply shows that the modules contained inside or overlapping the gray NoC ellipse are connected to the NoC. In Fig. 3.2, more information is given regarding the topology of the NoC and its use within the MORPHEUS platform. The dashed lines in Fig. 3.2 denote the interconnections in the NoC, whereas the boxes denote the NoC nodes. The NoC provides a routing mechanism that allows exchange of data between NoC nodes that are not adjacent (e.g. DREAM to XPP In transfer via ARM node).

To avoid a possible overload of certain NoC interconnections, assumptions were made about the expected communication behavior of the modules, which are connected to the NoC. The main idea for optimizing the topology is to place NoC nodes with high inter-communication demand directly adjacent to one another, since a direct interconnection link exists between such nodes. For example, since the XPP module is a typical data streaming device, the XPP NoC interfaces have been placed adjacent to the external memory controller. If a certain module has further demands for high bandwidth or low latency, more than one NoC interface node can be reserved for this module – provided the module can handle more than one communication interface. For example, in the NoC topology shown above, the

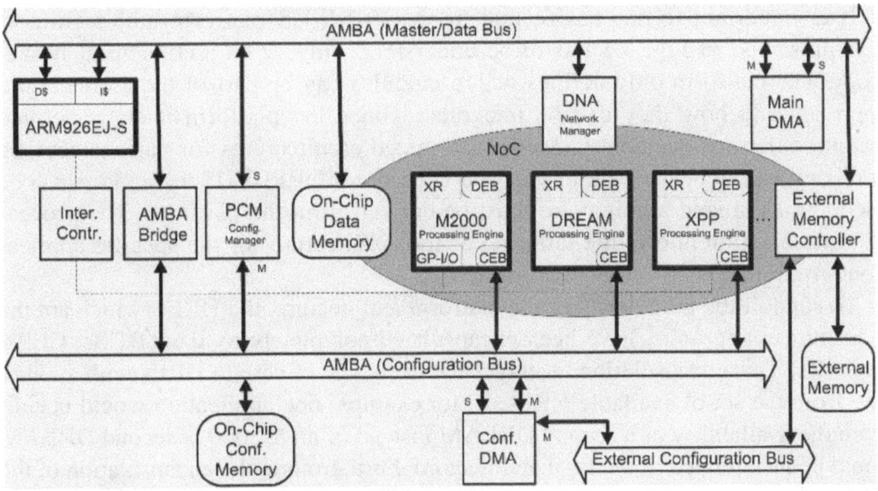

Fig. 3.1 Simplified MORPHEUS platform architecture

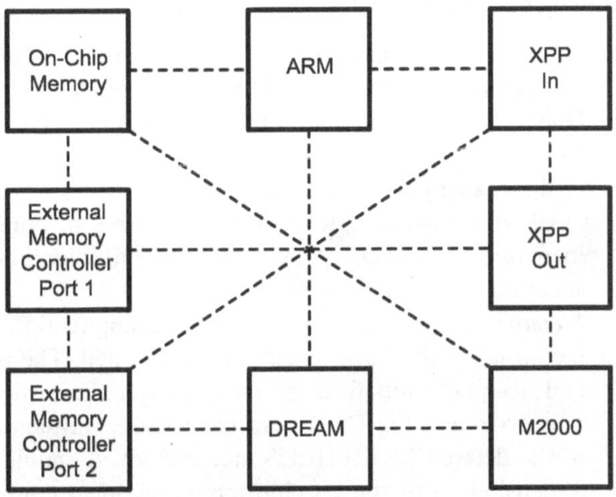

Fig. 3.2 MORPHEUS NoC topology

external memory controller is expected to supply very high data bandwidth. For this reason two NoC interface nodes are planned for this module.

3.4 Expandability of MORPHEUS Platform Architecture

One of the most attractive features of the MORPHEUS platform architecture is that it targets different application domains. Later on in this book, different application test cases will be presented, which have been used to evaluate the MORPHEUS approach.

It is important to emphasize that the MORPHEUS platform architecture as it is presented in Fig. 3.1 has to be understood only as an architectural framework. The platform only defines which modules can be part of the architectural approach and how they can be integrated. Since the platform itself is defined in a modular and scalable fashion, customized architectures for various specific applications can be derived. In every case, before MORPHEUS technology can be used, a customized architecture must be derived from the platform. The process of customization allows the tailoring of the architecture for the specific application requirements.

To support the modularity of the platform architecture, the HREs, which are the main processing units, have been encapsulated completely by their DEBs, CEBs, and XRs. This encapsulation leverages the exchange of certain HREs with another one from the set of available HREs. If, for example one application would benefit from the availability of a second DREAM instead of an M2000, a second DREAM could be instantiated in the final architecture. Furthermore, the encapsulation of the HREs also leverages the integration of new HREs into the platform architecture – even if this was not a major goal.

Scalability of the platform architecture is supported in multiple methods, some of which are mentioned in the following:

- *HREs*: The size and processing parameters of HREs, which characterize the computational power of the HREs is not predefined by the platform architecture. Due to the encapsulation of the HREs both can be adapted in the design process to the processing requirements of the application in focus. For example, the size of the coarse grain computation array in the XPP can be altered, or the internal clock speed of the DREAM and the sizes of configuration memories can be increased if necessary. Of course, such modifications have to be in line with the specifications of the respective HREs.
- *Memory subsystem*: Depending on the application requirements the size of all memories in the three levels can be adjusted. The platform architecture in general does not predefine any memory size. In particular the size of the on-chip memories and the DEBs can have strong influence on the final performance of the derived MORPHEUS architecture. It should also be mentioned that the dimensioning of the on-chip memories must consider potential limitations of the external memory data rate. Small on-chip memories may lead to an increased external memory bandwidth since intermediate results have to be stored in external memory. Finally, it was already mentioned that a custom designed external memory controller is recommended for the platform, but this controller is not obligatory [3].
- *NoC*: The MORPHEUS platform architecture integrates a NoC which is based on ST's spidergon technology with 8 nodes. From a conceptual point of view the number of NoC nodes is not limited and thus can be increased if necessary. However, it shall not be concealed that such an adaptation will have a huge impact on other architectural components such as the DNA controller or the Configuration Manager.

3.5 The MORPHEUS Prototype Chip

The concept of the presented MORPHEUS approach has been evaluated by derivation of one specific demonstration architecture from the platform architecture, as well as the production of a MORPHEUS prototype chip. Most of the evaluations of the MORPHEUS technology that are presented later in this book are based on this prototype chip.

It is obvious that the goals and limitations for such demonstration architectures are very different from the commercial and intended use of the MORPHEUS platform architecture, which usually focuses on a specific application. For the assessment of the MORPHEUS platform architecture it is important to distinguish whether possible limitations or disadvantages are based on the platform architecture itself or on one specific instantiation.

References

1. Coppola, M., Locatelli, R., Maruccia, G., Pieralisi, L., Scandurra, A., 2004, Spidergon: a novel on-chip communication network, *Proceedings of the International Symposium on System-on-Chip*, p.15ff, ISBN: 0-7803-8558-6.
2. Vassiliadis, S., Bertels, K., Kuzmanov, G., et al., 2004, The MOLEN polymorphic processor, *Proceedings IEEE Transactions on Computers*, Vol. 53, No. 11.
3. Whitty, S. and Ernst, R., 2008, A bandwidth optimized SDRAM controller for the MORPHEUS reconfigurable architecture, *Proceedings of the IEEE Parallel and Distributed Processing Symposium (IPDPS)* pp. 1–8, ISBN: 978-4244-1693-6.

Chapter 4
Flexeos Embedded FPGA Solution

Logic Reconfigurability on Silicon

Gabriele Pulini and David Hulance

Abstract This document describes the different features and architectural options of the fine grained M2000 eFPGA block for the MORPHEUS SoC.

Keywords FPGA Macro FlexEOS • embedded FPGA • eFPGA

4.1 Introduction

FlexEOS macros are SRAM-based, re-programmable logic cores using standard CMOS technology to be integrated into System on Chip (SoC) designs. FlexEOS is available in different capacities to achieve the required configurability while accommodating area and performance constraints. If necessary, multiple macro instances can be implemented in one device.

The macro core is delivered as a hard macro in a GDSII file, with all the necessary views and files required by a standard ASIC physical implementation flow.

This technology makes the MORPHEUS SoC re-configurable at any time during its life. The logic function of the core can be re-configured simply by downloading a new bitstream file. The embedded FPGA macro is used in the implementation of the MORPHEUS fine grained HRE as indicated in the Fig. 3.1 in Chapter 3.

4.2 Flexeos Macro Description

4.2.1 Overview of the FlexEOS Product

Each FlexEOS package contains the following items:

- A hard macro which is the actual re-configurable core to include in the SoC design.

G. Pulini (✉) and D. Hulance
M2000, France
gpulini@m2000.com

N. Voros et al. (eds.), *Dynamic System Reconfiguration in Heterogeneous Platforms*,
Lecture Notes in Electrical Engineering 40,

- A soft block which is the synthesizeable RTL description of the 'Loader', a controller which manages the interface between the macro core and rest of the SoC. Its main functions are to:
  – Load the configuration bitstream, and verify its integrity at any time.
  – Simplify the silicon test procedure.

Multiple macro instances in one SoC require multiple Loaders, one per macro.

- A comprehensive set of test patterns to test the macro in the SoC for production purposes
- A software tool suite to create:
  – Files required during the integration of the macro into the SoC design
  – A bitstream file to configure the hard macro for a particular application

4.2.2 FlexEOS Macro Block Diagram

Figure 4.1 shows a block diagram of a FlexEOS macro when embedded in an SoC, with the different interfaces to the rest of the system.

As can be seen in Fig. 4.1, each FlexEOS macro contains a macro core and a Loader.

The "Control Interface" is only used to access the system functions of the FlexEOS Macro, i.e. for writing commands and configuration words to the Loader and reading back status information from the macro core.

The "User Interface" signals correspond to the macro core input and output signals, and are the only ports which can be instantiated by a design mapped into the core during run-time.

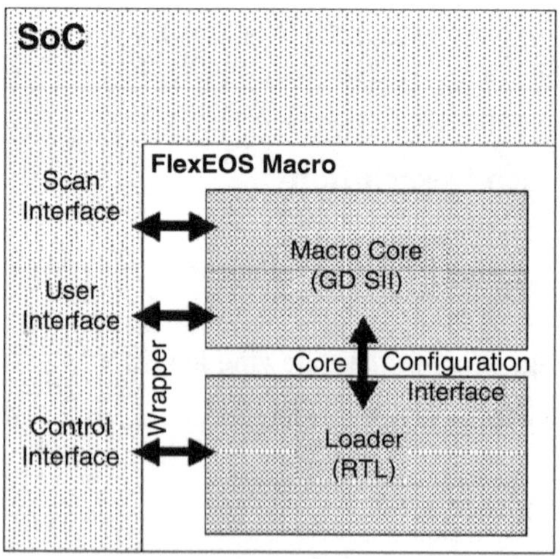

Fig. 4.1 FlexEOS Macro block diagram

4.2.3 Architecture

FlexEOS uses a highly scaleable architecture which permits gate capacities from a few thousands to multiple millions.

A possibility for the MORPHEUS SoC is the FlexEOS 4K macro which includes 4,096 MFCs (Multi-Function logic Cells).

4.2.3.1 The MFC

The basic FlexEOS building block is the Multi-Function logic Cell (MFC) which is a programmable structure with 7 inputs and 1 output. It combines a 4 input LUT (Look-Up Table) and a D flip-flop (Fig. 4.2).

The storage element has Clock, Clock Enable, and Reset input signals. The Clock signal always comes from the system clock tree, and can be inverted, whereas the Clock Enable and Reset signals can either come from the interconnect network via a regular signal input or from the system interconnect network. The FlexEOS compilation software selects the appropriate source according to the nature of the design to be implemented.

The MFCs are organized by groups of 16 and are all located at one hierarchical level in the core architecture.

4.2.3.2 Interconnect Network

FlexEOS eFPGA technology is based on a multi-level, hierarchical interconnect network which is a key differentiation factor in terms of density and performance when compared to other LUT-based FPGA technologies. The interconnect

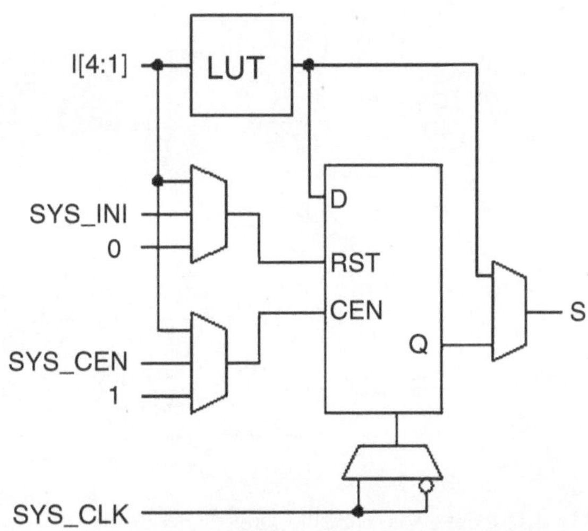

Fig. 4.2 MFC schematic

resources are based on a full crossbar switch concept (see Fig. 4.3), which provides equivalent routing properties to any element inside the macro and gives more freedom for placing and routing a given design to the FlexEOS compilation software.

Note that the interconnect network can only be configured statically.

Figure 4.4 shows the organization of the macro with the different building blocks. It also shows the symmetry of the architecture which provides more flexibility for mapping and placing a design. Each computing element of the macro can either be connected to its neighbor by using a local interconnect resource, or to another element via several interconnect resources.

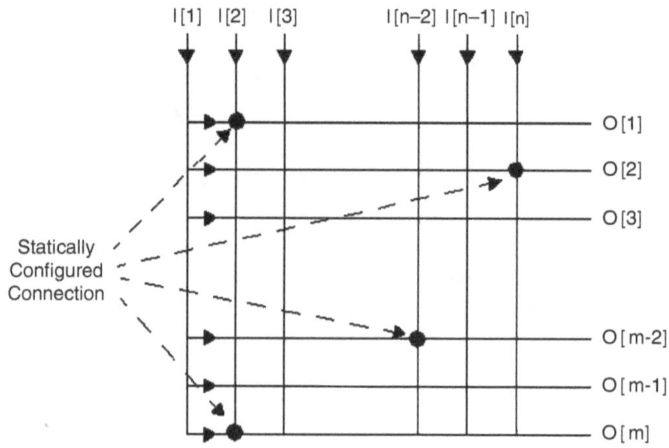

Fig. 4.3 Full crossbar switch

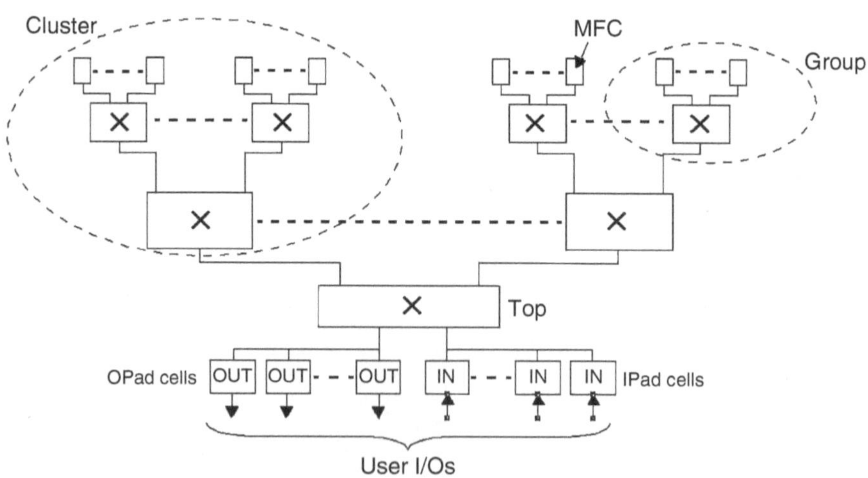

Fig. 4.4 FlexEOS core architecture

In addition to the regular interconnect network, a low-skew low-insertion-delay buffer tree network (system interconnect network) starts from eight dedicated user input ports (SYS_IN) and connects to all the synchronous cells. Its usage is recommended for high fanout signals such as reset signals, or high speed signals such as clock signals.

Note that if part of the system interconnect network is not used by the design, the FlexEOS compilation software automatically uses portions of it to improve the final design mapping and performance.

4.2.3.3 User I/O Interface

At any level of the hierarchy, the interconnect resources are unidirectional, including the user I/O interface signals. The standard 4K-MFC macro block includes 512 input ports and 512 output ports. Each of them is connected in the same way to the interconnect network, which gives the following properties:

- Any input port can access a given computing resource inside the core
- Any input port can be used as a system signal such as clock or reset
- Any output port can be reached by a computing resource

These three points are meaningful when considering the integration of the eFPGA macro inside the SoC architecture and defining the physical implementation constraints.

During the SoC design phase, several potential applications should be mapped to the eFPGA to:

- Evaluate the system constraints of the IP.
- Refine the different parameters of the IP (number of MFCs and I/Os, need for carry chains, memory blocks, MACs).
- Evaluate its connectivity to the rest of the system. This is made easier by the flexibility of the eFPGA interconnect network and its I/O ports properties: the FlexEOS macro does not add any routing constraints on SoC signals connected to the user I/Os as they can reach any resource inside the macro core.

Boundary Scan Chain

The core I/O cells are connected together internally to form two boundary scan chains:

- One for the input ports
- One for the output ports

They can be included in the SoC scan chains when implementing the chip to test the random logic connected the macro core I/Os. The boundary scan chain models are delivered as VHDL files and are compatible with standard ATPG tools.

4.2.3.4 Loader

The FlexEOS LUT-based FPGA technology needs to be configured each time the power is turned on, or to change its functionality. The macro is configured by a bitstream file which is handled by the Loader.

The design of the Loader is optimized to simplify interaction between the rest of the SoC and the macro core, and to allow predictable and reliable control of the core configuration and operation modes.

The Loader also verifies the integrity of the bitstream with a CRC check. The CRC signature computation cycle time is about 2 ms for a 4K-MFC macro, depending on the Loader clock frequency.

The Loader includes specific functions which speed up the silicon test time. It tests similar structures by simultaneously replicating a basic set of configuration and test vectors for the whole macro core. The results are stored in the Loader's status register which can be read by the external controller at the end of each test sequence to find out if it failed or passed.

The Loader is delivered as a synthesizable VHDL design, which requires between 10 and 20K ASIC gates, depending on the customer implementation flow and target manufacturing technology. Its typical operating frequency is in the 100 MHz range.

4.2.3.5 System Bus Interface

The system bus interface is directly connected to the FlexEOS Loader control interface. This interface behaves similarly to a synchronous SRAM block. It comprises the following signals (Fig. 4.5):

- Clock (100 MHz and below)
- Reset (active high), needs to be activated at power-on to reset the Loader and the core
- Data In (usually 32 bits, depending on the system bus width)

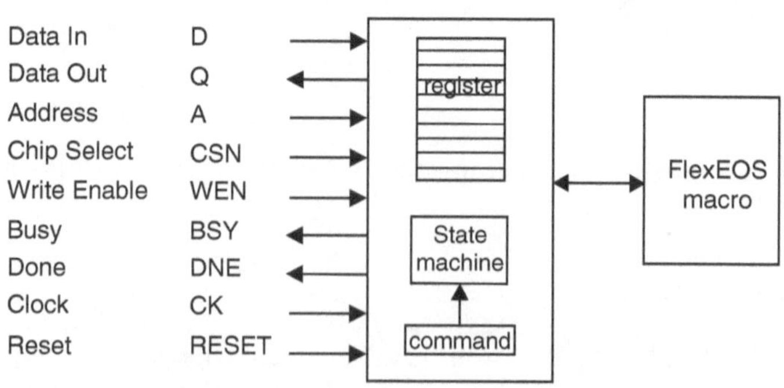

Fig. 4.5 FlexEOS loader overview

- Data Out (usually 32 bits, depending on the system bus width)
- Address (4 bits)
- Chip Select (active high)
- Write Enable (active high)
- Busy (active high)
- Done (active high)

A typical operation starts by writing a command and data to the appropriate registers. The state machine then executes the command, and sets the Busy signal to high. When the operation has completed, the Busy signal goes to low, and a subsequent command can be executed.

The eFPGA macro, together with its Loader, can be implemented multiple times on the chip, connecting to the system and/or peripheral busses.

4.2.4 Size and Technology

Table 4.1 shows the dimensions of a 4K FlexEOS macro in 90 nm CMOS technology with seven metal layers.

As an example, a FlexEOS macro with 4K MFCs has an equivalent ASIC gate capacity of up to 40,000 gates. The design configuration file (bitstream) size is 36Kbytes, and the loading time is around 600 μs range when the FlexEOS Loader operates at 100 MHz. The data bus interface is 32-bits wide.

Table 4.2 shows several examples of designs mapped to FlexEOS eFPGA macros. It also provides the correspondence between the ASIC gate count derived from Synopsys Design Compiler and the MFC capacity required to map the same designs to a FlexEOS macro. FlexEOS macros can be ported to any standard CMOS process. Multiple identical macros can be implemented on one SoC.

Table 4.1 FlexEOS 4K-MFC features and size

Equivalent ASIC gates	40,000 (Estimated when considering MFCs only)
LUTs/DFFs (MFCs)	4,096
I/Os	504 × IN, 512 × OUT, 8 × SYS_IN
Silicon area for 4K MFCs only	2.97 mm^2 (CMOS 90 nm)

Table 4.2 Example of design mapping results

	ASIC Gates	Equivalent MFCs (LUT + FF)	Typical CMOS 90 High-Vt	FlexEOS eFPGA macro sizes
160 × 16 bit counters	29,742	3,982	11.5 ns	4,096 MFCs
UART 16550	8,096	1,459	9.0 ns	1,536 MFCs
Viterbi Decoder	10,028	2,245	12.1 ns	3,072 MFCs
Ethernet MAC	20,587	3,995	~12 ns	4,096 MFCs

4.3 Soc Integration

M2000 provides the required files for assembling the SoC at the RTL level:

- The entity for the eFPGA hard-macro which comprises all the user input and output ports, as well as the system I/Os which are exclusively connected to the configuration interface IP.
- The Loader RTL (VHDL) IP which is the system interface for managing the configuration, operation modes and test. It can also be called the configuration interface IP.
- The top-level RTL wrapper which connects the Loader to the eFPGA macro and to the SoC system databus, and the eFPGA user IOs to the application signals.

As soon as the FlexEOS macro is integrated at the RTL level, the designer can start verifying that the access to the eFPGA system interface works correctly by simulating the configuration bitstream load operation.

M2000 can provide a full RTL model of the eFPGA hard-macro, or a simpler model to emulate the configuration hardware behavior. Such a simulation involves the following steps:

- Reset the Loader (external signal) and then the eFPGA macro (command sent to the Loader).
- Initialize the Loader to the proper operation mode (load, test, etc. … depending on the test-bench).
- Send the configuration bitstream data to the Loader. The data has to be transferred from a memory to the Loader bus interface through the AMBA to AHB bus interface.
- Activate the eFPGA macro by setting the Loader mode register to the proper value.

The designer now needs to run the application (such a simulation can be very slow if the full eFPGA model is being used).

After the designer has verified that the FlexEOS macro is correctly integrated, he can simulate an application by creating, synthesizing and mapping an RTL design into the eFPGA using the FlexEOS software flow.

4.3.1 FlexEOS Software Tool Suite

The FlexEOS proprietary software tool suite provides a design flow which is complete, easy to use, and designed to interface with the main standard FPGA synthesis software packages. It takes the following files as inputs:

- Design structural netlist mapped to DFFs and LUTs, generated by FPGA synthesis software.
- I/O pin assignment file, i.e. assignment of specific input or output I/O cells to each input and output port of the design.
- Design constraints such as clock definition, input and output timing delays, false path (see the FlexEOS compilation software documentation for more details).

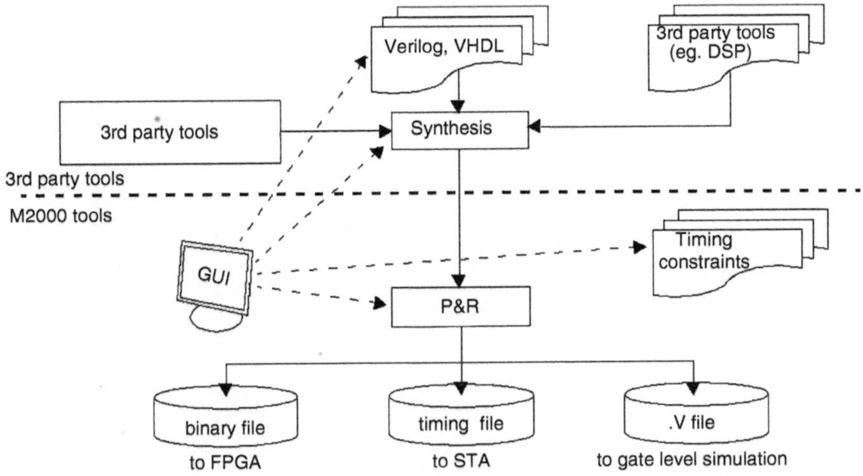

Fig. 4.6 FlexEOS software flow

The FlexEOS compilation software provides implementation options such as timing-driven place-and-route, automatic design constraint generation (very useful the first time a design is mapped).

The output files are:

- Configuration bitstream to be loaded in the eFPGA core.
- Configuration bitstream reference signature to be provided to the Loader.
- Functional Verilog netlist for post-implementation simulation.
- Timing annotation file (SDF: Standard Delay File) to perform further timing analysis on a given mapped design with third party software, or to run back-annotated simulation when used in combination with the generated Verilog netlist.
- Timing report for each clock domain critical path for a pre-selected corner (Best, Typical or Worst case).
- Macro wrapper (Verilog file) which instantiates the mapped design and connects its I/O ports to the physical core ports. This file is useful for in-context (i.e. in the SoC environment) timing analysis or simulation of applications.

The FlexEOS software flow is illustrated in Fig. 4.6. The RTL front-end design tasks are executed using commercial FPGA synthesis tools.

References

1. FlexEOS Software User Manual (on request from www.M2000.com).
2. The FlexEOS Loader Manual (on request from www.M2000.com).

Chapter 5
The Dream Digital Signal Processor

Architecture, Programming Model and Application Mapping

Claudio Mucci, Davide Rossi, Fabio Campi, Luca Ciccarelli, Matteo Pizzotti, Luca Perugini, Luca Vanzolini, Tommaso De Marco, and Massimiliano Innocenti

Abstract This chapter provides an overview of the DREAM digital signal processor. It discusses the programming model and the tool chain used to implement algorithm on the proposed architecture. It finally provides an application mapping example showing quantitative results.

Keywords Reconfigurable · mix-grain · datapath · PiCoGA · Griffy-C · PGA-OP

5.1 Introduction

Reconfigurable computing holds the promise of delivering ASIC-like performance while preserving run-time flexibility of processors. In many applications domains, the use of FPGAs [1] is limited by area, power and timing overhead. Coarse-Grain Reconfigurable architecture [2] offer computational density but at the price of being rather domain specific. Programmability is also a major issue related to all above described solutions.

A possible alternative that merges the advantages of FPGA-like devices and the flexibility of processor is the concept of Reconfigurable Processor [3], a device composed of a standard RISC processor that enables run-time instruction set extension on a programmable configurable hardware fabric.

This chapter describes the DREAM reconfigurable processor [4]. The computational core is composed by a mix grained 4-bit datapath allowing the signal processor to be suitable for a large set of applications, from error correction coding and CRC, to processing on binarized images. The design is completed with a full software tool-chain providing the application algorithmic analysis and design space exploration in an ANSI C environment using cycle accurate simulation and profiling.

C. Mucci, F. Campi, L. Ciccarelli, M. Pizzotti, L. Perugini, L. Vanzolini, and M. Innocenti
STMicroelectronics, Italy

D. Rossi (✉) and T.D. Marco
ARCES – University of Bologna, Italy

N. Voros et al. (eds.), *Dynamic System Reconfiguration in Heterogeneous Platforms*,
Lecture Notes in Electrical Engineering 40,

Its design strategy and generalized approach allow utilizing DREAM as a stand-alone IP in reconfigurable platforms controlled by state-of-the-art RISC processor. More precisely, the following design specifications were imposed: flexible homogeneous approach for IP integration, efficient utilization of local storage sub-system to optimize data communication with the host platform, user friendly programming model, well compliant with high-level languages.

5.2 Architecture Description

The DREAM digital signal processor is composed by three main entities: Control Unit, Memory Access Unit and a Configurable Datapath (PiCoGA).

Data transfers between DREAM and the host system are realized through communication buffers that also act as local data repository (Data Exchange Buffers, DEBs), program code/configuration storage (Configuration Exchange Buffers, CEBs).

DREAM features a local PLL, allowing to trade dynamically energy consumption with computation speed depending on the required data processing bandwidth, without any impact on the working frequency of the rest of the chip (Fig. 5.1).

5.2.1 *Control Unit*

The DREAM control unit fetches instructions, handles program flow, and provides appropriate control signals to the other blocks. Rather than utilizing a specific FSM, control tasks are mapped over a 32-bit RISC processor.

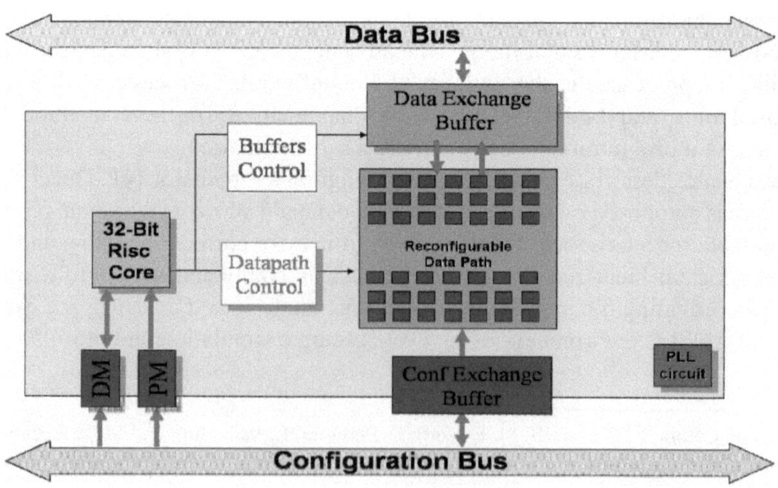

Fig. 5.1 DREAM architecture and interconnection with data and configuration buses of the host system. PM = Program Memory, DM = Data memory containing code instruction and data for the embedded processor. Computation data, on the contrary, is stored in the Data Exchange Buffers

Synchronization and communication between the IP and the host system main processor is ensured by asynchronous interrupts on the local core, and a cross-domain control register file (XRs). Processor code and data, as well as the embedded datapath configuration bitstream are considered as part of the DREAM program code, and are loaded by the host system on the Configuration Exchange Buffers (CEBs), implemented on dual port, dual clock memories. Memory sizes are configurable at HDL compilation time, and in the implementation here described are composed by 4K + 4Kbytes of processor code and data memory, plus 36Kbytes of datapath configuration memory. Input data and computation results are exchanged through a coarse-grained handshake mechanism on DEBs (also defined ping-pong buffering).

The choice of utilizing a small processor allows the user to exploit a sophisticated program control flow mechanism, writing commands in ANSI-C and utilizing a reliable compiler to optimize code and schedule task efficiently. The processor function units can also act as computation engines in some cases, concurrently to the reconfigurable datapath.

Computation kernels are re-written as a library of *macro-instructions*, and mapped on the reconfigurable engine as concurrent, pipelined function units. Computation is handled by the RISC core in a fashion similar to the *Molen* paradigm [5]: the core explicitly triggers the configuration of a given macro-instruction over a specific region of the datapath, and when the loading of the configuration is complete it may run any desired issue of the same functionality in a pipelined pattern.

Up to four macro-instructions can be loaded on each of the four available contexts. Contexts can not be computed concurrently but context switch requires only one cycle. A sophisticated stall and control mechanism ensures that only correctly configured operations can be computed on the array, and manages context switches.

5.2.2 Data Storage and Memory Access Architecture

In order to allow DREAM to function at the ideal frequency, regardless limitations imposed by the host system, dual clock embedded memory cuts were chosen as physical support for DEBs and CEBs. This caused a 5% overhead in timing, 40% in area and 20% in power consumption. This price is justified by the absence of multiplexing logic that would be required by the use of single port memories. This choice also offers a very straightforward physical implementation of the overall system, without need for explicit synchronization mechanisms that would require additional standard cell area and careful asynchronous timing evaluation in the back-end process. DEBs are composed by 16 dual port banks of 4Kbytes each. They are accessed as a single 32-bit memory device from the system side, but they can provide concurrent 16×32 bit bandwidth to/from the datapath (Fig. 5.2).

On the reconfigurable datapath side, an address generator (AG) is connected to each bank. Address Generation parameters are set by specific control instructions,

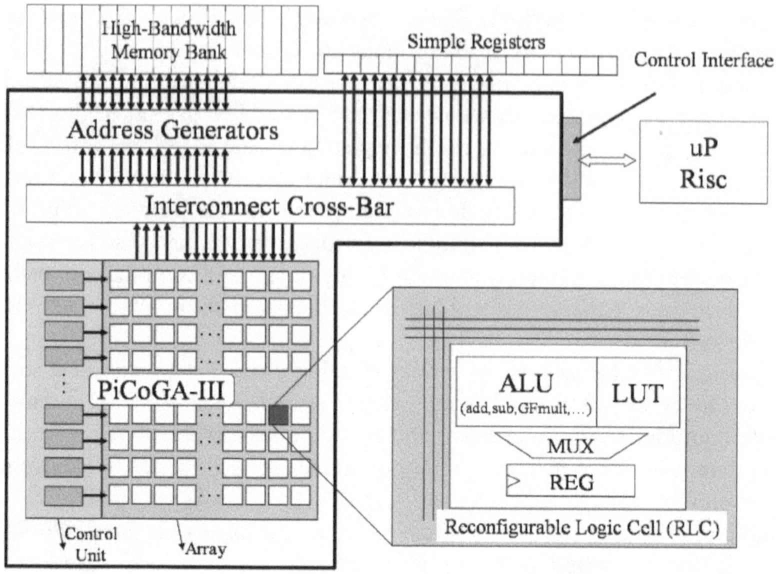

Fig. 5.2 DREAM architecture

and addresses are then incremented automatically at each cycle for all the duration of the kernel. AGs provide standard STEP and STRIDE [6] capabilities to achieve non-continuous vectorized addressing.

A specific MASK functionality also allows power-of-2 modulo addressing in order to realize variable size circular buffers with programmable start point.

Due to their small granularity, DREAM macro-instructions often exchange information between successive issues, in form of temporary results or control information. For this reason a specific 16-registers multi-ported register file is available as local data repository.

5.2.3 *PiCoGA*

The PiCoGA [7] is a programmable gate array especially designed to implement high-performance algorithms described in C language. The focus of the PiCoGA is to exploit the Instruction Level Parallelism (ILP) present in the innermost loops of a wide spectrum of applications (e.g. multimedia, telecommunication and data encryption).

From a structural point of view is composed by an array of Reconfigurable Logic Cells (RLCs). Each cell may compute 2 4-bit inputs and provide a 4-bit result. It is composed of a 64 bit LUT, a 4-bit ALU, a 4-bit multiplier slice and a Galois Field multiplier over GF(24). A carry chain logic is provided row-wide allowing fast 8-, 16- and 32-bit arithmetics.

The ideal balancing between the need for high parallelism and the severe constraints in size and energy consumption suggested a size of 16 × 24 RLCs, and an IO bandwidth of 384 inputs (1,232-bit words) inputs and 128 outputs (432-bit words).

The routing architecture features a 2-bit granularity, and is organized at three levels of hierarchy: global vertical lines carry only datapath input and outputs, while horizontal global lines may transfer temporary signals (i.e. implementing shifts without logic occupation). Local segmented lines (3 RLC per segment) handle local routing, while direct local connections are available between neighbouring cells belonging to the same column.

The gate-array is coupled to an embedded programmable control unit, that provides synchronous computation enable signals to each row, or set of rows of the array, in order to provide a pipelined data-flow according to data dependencies in the source DFG. Figure 4.2 shows an example of pipelined DFG mapped onto PiCoGA.

Due to its medium-grain and multi-context structure the DREAM datapath provides a good trade-off between gate density (3Kgates/mm^2 per each context) and flexibility. Its heavily pipelined nature allows a very significant resource utilization ratio (more than 50% of available resources are utilized per clock on average) with respect to devices such as embedded FPGAs that need to map on reconfigurable fabrics the control logic of the algorithm.

The full configuration of each context of the array is composed by 2Kbytes, that can be loaded in 300 cycles, but each operation can be loaded and erased from the datapath separately. To achieve this goal, the reconfigurable unit is organized in four contexts; one context can be programmed while a second one is computing. An on-board configuration cache (36Kbytes in the current implementation) and an high bandwidth configuration bus (288 bit per clock) are used in order to hide the reconfiguration process of one context in the time consumed by computation on different contexts.

Summarizing, with respect to a traditional embedded FPGAs featuring an homogeneous island-style architecture, the PiCoGA is composed of three main sub-parts, highlighted in Fig. 5.3:

- An homogeneous array of 16×24 RLCs with 4-bit granularity (capable of performing operations e.g. between two 4-bitwise variables) and connected through a switch-based 2-bitwise interconnect matrix
- A dedicated Control Unit which is responsible to enable the execution of RLCs under a dataflow paradigm
- A PiCoGA Interface which handles the communication from and to the system (e.g. data availability, stall generation, etc.)

5.3 Programming Approach

The language used to configure the PiCoGA in order to efficiently implement pipelined DFG is called Griffy-C [8]. Griffy-C is based on a restricted subset of ANSI C syntax enhanced with some extensions to handle variable resizing and register

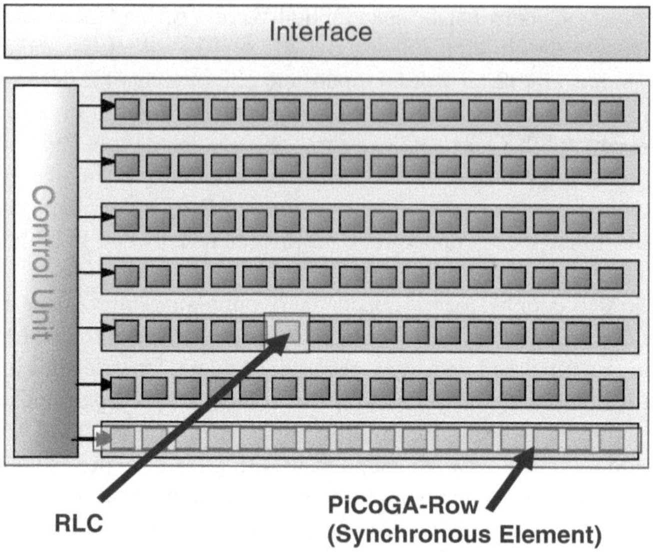

Fig. 5.3 Simplified PiCoGA architecture

allocation inside the PiCoGA: differences with other approaches reside primarily in the fact that Griffy is aimed at the extraction of a pipelined DFG from standard C to be mapped over a gate-array that is also pipelined by explicit stage enable signals.

The fundamental feature of Griffy-based algorithm implementation is that Data Flow Control is not synthesized on the array cells but it is handled separately by the hardwired control unit, thus allowing a much smaller resource utilization and easing the mapping phase. This also greatly enhances the placing regularity.

Griffy-C is used as a friendly format in order to configure the PiCoGA using hand-written behavioural descriptions of DFGs, but can also be used as an intermediate representation automatically generated from high-level compilers. It is thus possible to provide different entry points for the compiling flow: high-level C descriptions, pre-processed by compiler front-end into Griffy-C, behavioural descriptions (using hand-written Griffy-C) and gate level descriptions, obtained by logical synthesis and again described at LUT level. Restrictions essentially refer to supported operators (only operators that are significant and can benefit from hardware implementation are supported) and semantic rules introduced to simplify the mapping into the gate-array. Three basic hypotheses are assumed:

- *DFG-based description*: no control flow statements (if, loops or function calls) are supported, as data flow control is managed by the embedded control unit.
- *Single assignment*: each variable is assigned only once, avoiding hardware connection ambiguity.
- *Manual dismantling*: only single operator expressions are allowed (similarly to intermediate representation or assembly code).

Table 5.1 Basic operations in Griffy-C

<table>
<tr><td>Arithmetical operators</td></tr>
<tr><td>Dest = src1 [+, −] src2;</td></tr>
<tr><td>Bitwise logical operators</td></tr>
<tr><td>Dest = src1 [&,|,∧] src2;</td></tr>
<tr><td>Shift operators</td></tr>
<tr><td>Dest = src1 [<<, >>] constant;</td></tr>
<tr><td>Comparison operators</td></tr>
<tr><td>dest = src1 [<,<=,==,!=,>=,>]</td></tr>
<tr><td>Conditional Assignment (Multiplexer operator)</td></tr>
<tr><td>dest = src1 ? src2:src3</td></tr>
<tr><td>Extra-C operators</td></tr>
<tr><td>LUT operator:dest = src1@ 0x[LUT layout]</td></tr>
<tr><td>Concatenation operator:dest = src1 # src2</td></tr>
</table>

Basic Griffy-C operators are summarized in Table 5.1, while special intrinsic functions are provided in the Griffy-C environment in order to allow the user to instance non-standard operations.

Native supported variable types are signed/unsigned int (32-bit), short int (16-bit) and char (8-bit). Width of variables can be defined at bit level using #pragma directives. Operator width is automatically derived from the operand sizes. Variables defined as static are used to allocate static registers inside the PiCoGA, which is registers whose value is maintained across successive PGAOP calls (i.e. to implement accumulations). All other variables are considered "local" to the operation and are not visible to successive PGAOP calls.

Once critical computation kernels are identified through a code profiling step in the source code, they are rewritten using Griffy-C and can be included in the original C sources as atomic PiCoGA operations. #pragma PiCoGA directives are used to retarget the compiling flow from standard assembly code to the reconfigurable device as shown in Fig. 5.4.

Hardware configuration for elementary operation is obtained by direct mapping of predefined Griffy-C library operators. Thanks to this library-based approach, specific gate-array resources can be exploited for special calculations, such as a fast carry chain, in order to efficiently implement arithmetic or comparison operators. Logic synthesis is kept to a minimum, implementing only constant folding (and propagation) and routing-only operand extraction such as constant shifts: those operations are implemented collapsing constants into destination cells, as library macros have soft-boundaries and can be manipulated during the synthesis process.

Once a Griffy-C description of a DFG has been developed the automated synthesis tools (Griffy-C compiler) are then used to:

1. Analyze all elementary operations described in the Griffy-C code composing the DFG, determining the bit-width and their dependencies. Elementary operations are also called DFG *nodes*.

```
#pragma pga SAD4 1 2 out p1 p2
{

  unsigned char p10, p11, p12, p13;
  unsigned char cond0, cond1, cond2, cond3;

                ...

  #pragma attrib cond0, cond1, cond2, cond3 SIZE=1

                ...

  sub0a = p10 - p20;  sub0b = p20 - p10;
  cond0 = sub0a < 0;
  sub0  = cond0 ? sub0b : sub0a;

                ...

  out = acc1 + acc2;

}
#pragma end
```

Fig. 5.4 Example of Griffy-C code representing a SAD (sum of absolute differences)

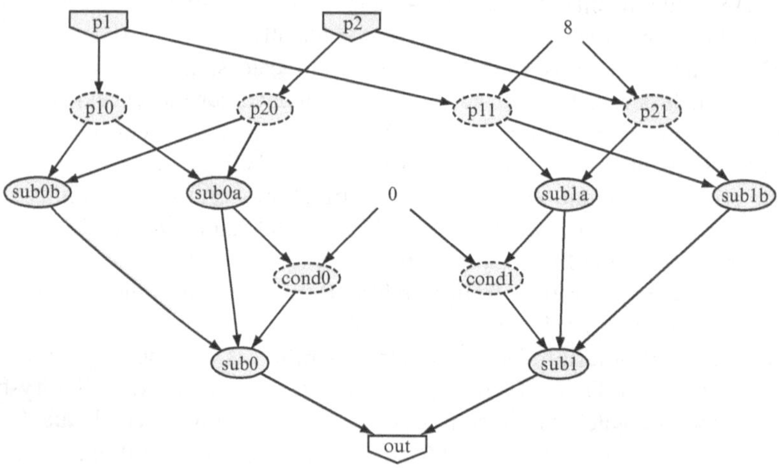

Fig. 5.5 Example of pipelined DFG

2. Determine the intrinsic ILP between operations (nodes). Figure 5.5 shows an example of pipelined DFG automatically extracted from a Griffy-C description. In this representation, nodes are aligned for pipeline stage.
3. Map the logic operands on the hardware resources of the PiCoGA cells (a cell is formed by a Lookup Table, an ALU, and some additional multiplexing and computational logic). Each cell features a register that is used to implement pipelined computation. Operations *cannot be cascaded over two different rows.*

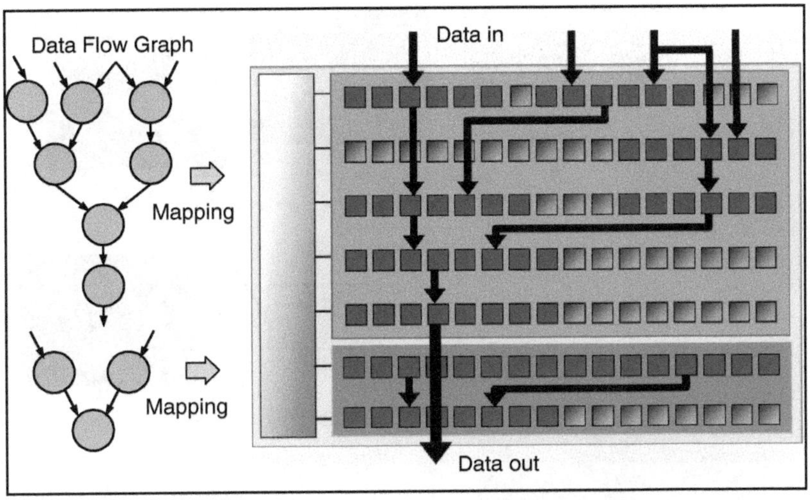

Fig. 5.6 DFG mapping on PiCoGA

4. Route the required interconnections between RLCs using the PiCoGA interconnection channels.
5. Provide the bitstream (in the form of a C vector) to be loaded in the PiCoGA in order to configure both the array and the control unit (the PiCoGA Interface does not require a specific configuration bitstream). Configurations are relocatable, thus they can be loaded in any configuration layer starting from any available row.

Figure 5.6 represents a typical example of mapping onto PiCoGA. As explained in previous sections, after a data-dependency analysis, the DFG is arranged in a set of pipeline stages (thus obtaining the pipelined DFG). Each of the pipeline stages is placed in a set of rows (typically they are contiguous rows, but this is not mandatory). In Fig. 5.6 different rows represent different pipeline stages. Depending on the row-level granularity of the PiCoGA Control Unit, one row can be assigned only to one single pipeline stage, and it cannot be shared among different pipeline stages.

5.4 Application Mapping Example: Motion Detection Algorithm

The Motion detection algorithms provides the capability to detect a human, a vehicle or an object in movement with respect to a static background. This could be useful for example to activate an alarm in case of security applications or to start the recording in case of area monitoring system. A typical motion detection algorithm is shown in Fig. 5.7.

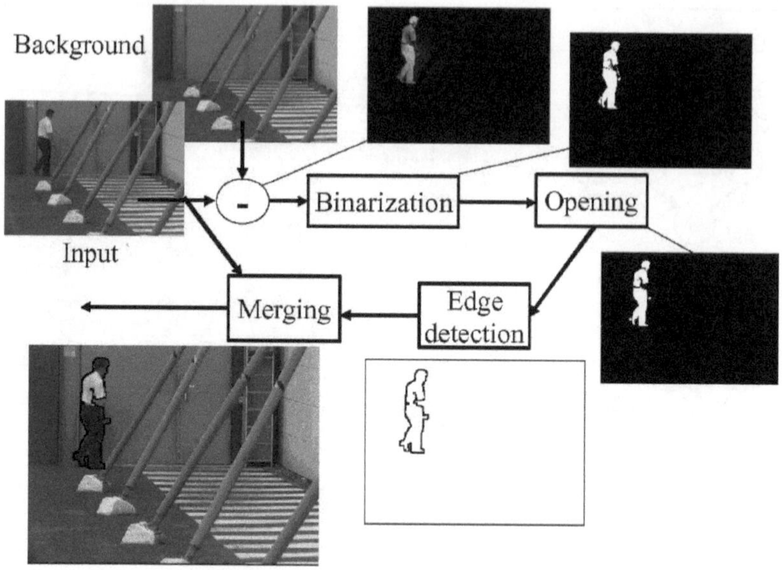

Fig. 5.7 Simplified motion detection algorithm overview

Most of the processing is performed on the image resulting from the absolute *pixel-to-pixel difference* between the current frame and the background, which can be stored during the setup of the intelligent camera. Even if this differentiation allows to isolate the object under analysis (if any), too many details are present as the complete grayscale. For that, *binarization* is applied to the frame. Given a threshold conventionally fixed to 0.3 times the maximum value of the pixels, binarization returns a Top Value if the current pixel is greater than the threshold, and a Bottom Value otherwise. The resulting image could still be affected from noise (spurious pixels) or on the contrary could be affected by some hole. This "cleaning" task is accomplished by the opening phase, implemented by two operators:

- **Erosion**, that working on 3×3 pixel matrices substitutes the central pixel with the minimum in the matrix, thus removing random noise
- **Dilatation**, that working on 3×3 pixel matrices substitutes the central pixel with the maximum in the matrix, closing eventual small holes and reinforcing details

The next step is the *edge detection* which allows identification of the boundaries of the human or object that is moving in the monitored area. This operation is implemented by a simple convolution applied to 3×3 pixel matrices using the Sobel algorithm [9]. The resulting image is then *binarized*, since the aim of the application is not to detect the magnitude of the gradient but the presence of a gradient. Finally the detected edge is merged with the original image. For that goal, inverse binarization is applied: the background is filled by 1s and moving image edges by 0s, thus allowing to implement the merge operation with a multiplication.

5.4.1 Motion Detection on DREAM

The implementation or more in general the acceleration of the above described application on the reconfigurable platform, is driven by two main factors:

- *Instruction/data level parallelism*: each operation shows relevant instruction level parallelism. Given a specific image processing kernel, computation associated to each pixel is independent from the elaboration of other pixels, although the reuse of adjacent pixel proves beneficial to minimize memory access.
- *Data size*: after binarization, the information content associated to each pixel can be represented by only 1 bit (edge/no edge), thus allowing to store up to 32 pixel in a 32-bit word. This significantly reduces memory utilization without implying additional packing/unpacking overhead, as it would be the case with 32-bit processors, since DREAM may handle shifts by programmable routing.

This last consideration provides additional benefits since:

1. **Erosion** phase requires the search of the minimum within the pixels in a 3×3 matrix, but is implemented on DREAM by a single 9-bit input 1-bit output AND.
2. **Dilatation** phase requires the search of the maximum within the pixels in a 3×3 matrix, but is implemented on DREAM by a single 9-bit input 1-bit output OR.
3. **Edge detection** requires to detect the presence of a gradient. The Sobel convolution is implemented on DREAM using 4-bit LUTs. Since the required information is not a magnitude but the *presence* of a gradient, the final binarization can be achieved DREAM by 2 8-input NOR for each pixel, one for the vertical convolution and one for the horizontal convolution.
4. **Final merging** phase can be implemented on DREAM as an 8-bit bitwise AND operation, instead of an 8-bit multiplication as a consequence of the edge detection simplification.

The processing chain is based on simple operations repeated many times for all the pixels in a frame. In the case of DREAM and PiCoGA, all these operations do not fully exploit parallelism made available by the reconfigurable device. It is thus possible to operate concurrently on more pixels at time unrolling inner loops in the computation flow.

We concurrently compute on three different rows at a time for all the computations, since most of the operations requires pixels from three adjacent rows:

- **Erosion**, **dilatation** and **edge detection** read data from three adjacent rows and provide the result for the row in the middle. In this case, since each pixel is represented by 1 bit, we elaborate $3 \times 32 = 96$ pixels per PiCoGA operation, packing 32 pixels in a single 32-bit memory word stored in the local buffer.
- **The other operation** can work on three adjacent rows to maintain a certain degree of regularity in the data organization. In this case, 8 bits are used to represent a pixel and we can pack 4 pixel in each memory word, resulting in an elaboration of 12 pixel at time.

To allow the concurrent access to three adjacent rows, we use a simple 3-way interleaving scheme in which each row is associated to a specific buffer by the rule: *buffer_index = row mod 3*. Rows are stored contiguously in each buffer, and each PiCoGA operation read a row chunk per cycle.

Address generators are programmed in order to scan the buffers where rows are stored according to the above described access pattern, while programmable interconnect is used to dynamically switch between referenced rows.

Boundary effects due to the chunking are handled internally to PiCoGA that can hold the pixels required for the different column elaborations in internal registers, thus avoiding data re-read.

Depending on the size of the frame, the available level of pipelining increases, augmenting the number of pixels in a row. As a consequence, the amount of memory required to store the frame under elaboration also increases. Bigger frames can be elaborated in chunks, performing the computation on sub-frames.

5.4.2 Application Mapping Results

Table 5.2 shows the cycle-count in the case of 80×60 chunks. This is an interesting sub-case since all necessary data for computation can be stored locally to DREAM, including both the original frame and the background which are the most demanding contributions in term of memory requirements. Cycle-counts are reported normalized with respect to the number of pixels in the image, to give a direct comparison with respect to the pure software implementation in Table 5.2.

Figure 5.8a shows the potential performance gain for bigger frames in terms of cycles/pixel reduction. Speedups range from 342× times to ~1,200×. It should be noted that larger images cannot be stored entirely on the DREAM memory sub-system, although the packetization performed after binarization allows to hold internally the most critical part of the computation up to the frame size 640 × 480. Figure 5.8b shows the dramatic reduction in memory access provided by the DREAM solution. For the software solution, memory transfers are considered for pixel access only, without including accesses due to temporary results, stack management and so on. Considering the overall motion detection application, the

Table 5.2 DREAM-based implementation results for an 80 × 60 Chunk

Kernel	Cycles/pixel	Speedup
Absolute Difference	0.11	173×
Max pixel value	0.09	100×
Binarization	0.45	24×
Erosion	0.42	326×
Dilatation	0.42	326×
Inv. Bin. Edge Detection	0.43	914×
Merging	0.17	59×
Total	2.09	342×

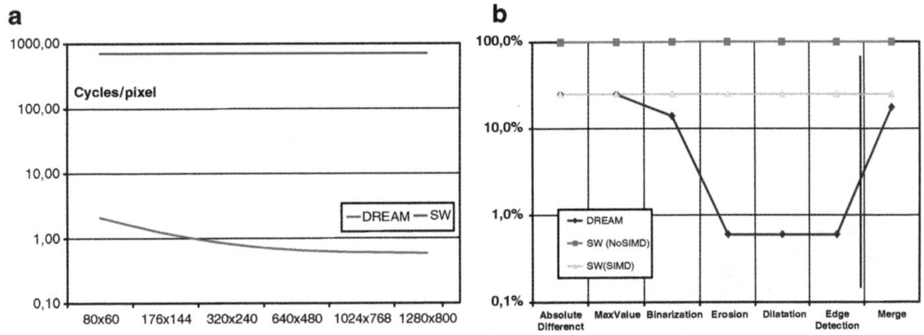

Fig. 5.8 (**a**) Performance gain vs. frame size. (**b**) Normalized memory access per pixel

DREAM solution needs roughly 2 memory accesses per pixel, whereas a software implementation requires ~39 memory accesses per pixel. We also considered a software-optimized SIMD-like access in which the processor could be able to handle 4-pixel per cycle (also during comparison). Also in this case our reconfigurable solution achieves ~80% memory access reduction, with a consequent benefit in terms of energy consumption.

References

1. R. W. Hartenstein, A decade of reconfigurable computing: a visionary retrospective, *Proceedings* <http://ieeexplore.ieee.org/xpl/RecentConjsp?punumber=7307>of *DATE*, 642–649, Mar. 2001.
2. A. De Hon, The density advantage of Configurable Computing, *IEEE Computer*, 33(4), 41–49, Apr. 2000.
3. J. Nurmi, Processor Design: System-on-Chip Computing for ASICs and FPGAs, Chapter 9, 177–208, Apr. 2007.
4. F. Campi, A. Deledda, M. Pizzotti, L. Ciccarelli, C. Mucci, A. Lodi, A. Vitkovski, L. Vanzolini, P. Rolandi, A dynamically adaptive DSP for heterogeneous reconfigurable platforms, *Proceedings on IEEE/ACM DATE*, 1–6, Apr. 2007.
5. S. Vassiliadis, S. Wong, G. Gaydadjiev, K. Bertels, G. Kuzmanov, E. M. Panainte, The MOLEN Polymorphic Processor, *IEEE Transaction on Computers*, pp. 1363–1375, Nov. 2004.
6. S. Ciricescu, R. Essick, B. Lucas, P. May, K. Moat, J. Norris, M. Schuette, and A. Saidi, The reconfigurable streaming vector processor, *ACM International Symposium on Microarchitecture*, pp. 141–150, Dec. 2003.
7. A. Lodi, C. Mucci, M. Bocchi, A. Cappelli, M. De Dominicis, L. Ciccarelli, A multi-context pipelined array for embedded systems, *International Conference on Field Programmable Logic and Applications, 2006. FPL'06*, pp. 1–8, Aug. 2006.
8. C. Mucci, C. Chiesa, A. Lodi, M. Toma, F. Campi, A C-based algorithm development flow for a reconfigurable processor architecture, *IEEE International Symposium on System on Chip*, pp. 69–73, Nov. 2003.
9. C. Mucci, L. Vanzolini, A. Deledda, F. Campi, G. Gaillat, Intelligent cameras and embedded reconfigurable computing: a case-study on motion detection, *International Symposium on System-on-Chip*, pp. 1–4, Nov. 2007.

Chapter 6
XPP-III

The XPP-III Reconfigurable Processor Core

Eberhard Schüler and Markus Weinhardt

Abstract XPP-III is a fully programmable coarse-grain reconfigurable processor. It is scalable and built from several modules: the reconfigurable XPP Array for high bandwidth dataflow processing, the Function-PAEs for sequential code sections and other modules for data communication and storage. XPP-III is programmable in C and comes with a cycle-accurate simulator and a complete development environment. A specific XPP-III hardware implementation is integrated in the MORPHEUS chip.

Keywords Coarse-grain reconfigurable · reconfiguration · dataflow · control-flow · VLIW core · XPP Array · FNC-PAE

6.1 Introduction

The limitations of conventional sequential processors are becoming more and more evident. The growing importance of stream-based applications makes coarse-grain reconfigurable architectures an attractive alternative. They combine the performance of ASICs with the flexibility of programmable processors. On the other hand, irregular control-flow dominated algorithms require high-performance sequential processor kernels for embedded applications.

The XPP-III (*eXtreme Processor Platform III*) architecture combines both, VLIW-like sequential processor kernels optimized for control-flow dominated algorithms and a coarse-grain reconfigurable dataflow array (*XPP Array*) for data streaming applications. XPP-III is designed to support different types of parallelism: pipelining, instruction level, dataflow, and task level parallelism. Additionally, a high-bandwidth communication and memory access framework (which is an integrated part of the XPP-III architecture) provides the performance and flexibility to feed the parallel XPP-III processing kernels with data and to integrate it into any SoC.

E. Schüler (✉) and M. Weinhardt
PACT XPP Technologies, Germany
eberhard.schueler@pactxpp.com

N. Voros et al. (eds.), *Dynamic System Reconfiguration in Heterogeneous Platforms*, Lecture Notes in Electrical Engineering 40,

XPP-III meets the performance requirements of heterogeneous embedded applications and accelerators. It is well suited for applications in multimedia, media streaming servers, telecommunications, simulation, digital signal processing, cryptography and similar application domains. The XPP-III architecture is highly scalable and enables adaptation to any application-driven chip specification.

XPP-III includes a set of *XPP-III components* which allow the SoC designer to assemble the final IP very easily. This has been demonstrated in the MORPHEUS project where the XPP-HRE was designed from standard XPP-III components and a small number of interfaces which provide the link to the top-level MORPHEUS SoC architecture, e.g. to the Network on Chip (NoC) and to the AMBA Bus.

The following sections give an overview of the XPP-III features and programming. For more details on XPP-III, refer to the XPP White Papers [1–4].

6.2 XPP Basic Communication Mechanism

The basic communication concept of the XPP-III architecture is based on streams. On the XPP architecture, a *data stream* is a sequence of single data packets traveling through the flow graph that describes the algorithm. A data packet is a single machine word (16 bit in the MORPHEUS implementation). Streams can, e.g., originate from external streaming sources like A/D converters or from memory (via DMA controllers or NoC). Similarly, data computed by the XPP can be sent to streaming destinations such as D/A converters, internal or external RAMs.

In addition to *data* packets, state information packets ("events") are transmitted through independent event connections. Event packets contain one bit of information and are used to control the execution of the processing nodes and may synchronize external devices.

The unique XPP communication network enables automatic synchronization of packets. An XPP object (e.g. an ALU) operates and produces an output packet only when all input data and event packets are available. The benefit of the resulting self-synchronizing network is that only the number and order of packets traveling through a graph is important. There is no need for the programmer or compiler to care about absolute timing of the pipelines during operation. This hardware feature provides an important abstraction layer allowing compilers to effectively map programs to the array.

6.3 XPP Dataflow Array

The XPP Array (Fig. 6.1) is built from a rectangular array of two types of *Processing Array Elements* (*PAE*): Those in the center of the array are *ALU-PAEs*. At the left and right side of the ALU-PAEs are *RAM-PAEs with I/O*. An ALU-PAE contains three 16-bit ALUs, two in top-down direction and one in bottom-up direction.

A RAM-PAE contains two ALUs, a small RAM and an I/O object. The I/O objects provide access to external streaming data sources or destinations.

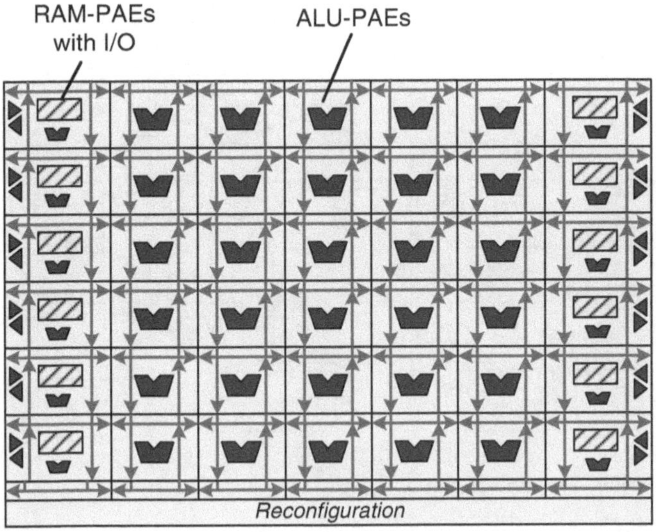

Fig. 6.1 XPP-III dataflow array

The XPP's data and event synchronization mechanism is extended to the I/O ports by means of handshake signals.

The horizontal routing busses for point-to-point connections between XPP objects (ALUs, RAMs, I/O objects, etc.) are also integrated in the PAEs. Separate busses for 16-bit data values and 1-bit events are available. Furthermore, vertical routing connections are provided within the ALU-PAEs and RAM-PAEs.

The real strength of the XPP Array originates from the combination of parallel array processing with fast run-time reconfiguration mechanisms [2,5]. PAEs can be configured while neighboring PAEs are processing data. Entire algorithms can be configured and run independently on different parts of the array. Reconfiguration is triggered by an external processor like a Function-PAE with the help of the Configuration DMA controller. A reconfiguration typically requires only a few thousand cycles. This is several orders of magnitude faster than FPGA reconfiguration.

6.4 Function-PAE

Control-flow dominated, irregular and strictly sequential code is mapped to one or several concurrently executing *Function-PAEs* (FNC-PAEs). They are sequential 16-bit processors which are optimized for sequential algorithms requiring a large amount of conditions and branches like bit-stream decoding or encryption.

FNC-PAEs (Fig. 6.2) are Harvard processors and similar to VLIW DSPs. But they provide more flexibility and unique features. A FNC-PAE executes up to eight ALU operations and one side function (in the *Side Function Unit*, *SFU*) in one

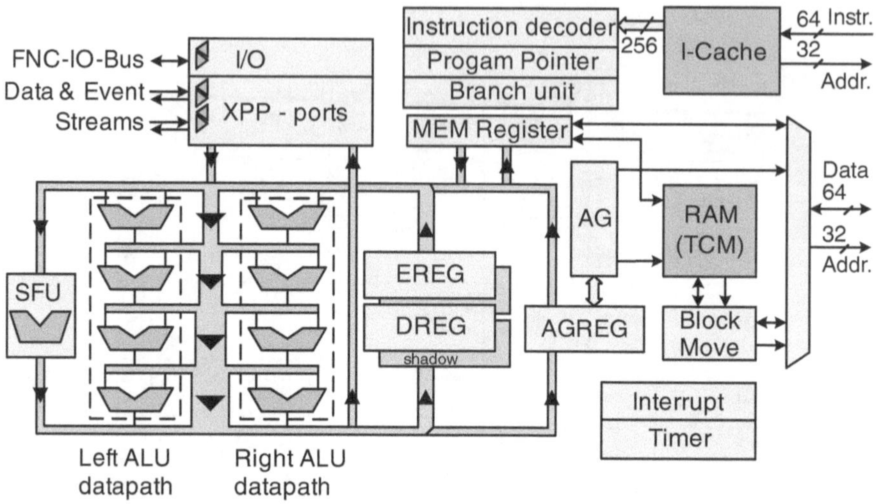

Fig. 6.2 FNC-PAE structure

clock cycle. Operations on up to four ALU levels can be *chained*, i.e. the output of one operation is immediately fed to the input of the next operation in the chain. This can even be combined with *predicated execution*, i.e. conditional execution based on the results of input operations. In this way, nested if-then-else statements can be executed in one cycle. Furthermore, special mechanisms enable jumps (conditional or non-conditional) in one cycle.

The eight ALUs are designed to be small and fast because they are arranged in two combinational columns of four ALUs each. The ALUs are restricted to a limited instruction set containing arithmetic, logic, comparison and barrel shift operations including conditional execution and branching. The benefit of the combinatorial data path is that the clock frequency can be reduced in order to save power.

Every ALU selects its operands from the register files (data registers *DREG* and *EREG*, both with shadow registers), the address generator registers (*AGREG*), the memory register *MEM*, or the ALU outputs of all rows above itself. Furthermore, the ALUs have access to I/O ports. All ALUs can store their results simultaneously to the registers.

The ALU datapath is not pipelined since the FNC-PAE is optimized for irregular code with many conditions and jumps. These code characteristics would continuously stall the operator pipeline, resulting in a low IPC (instructions per cycle) count. Instead, the FNC-PAE chains the ALUs and executes all instructions asynchronously in one cycle, even if there are dependences. Together with unique features which enhance the condition execution and branching performance, this results in a very high IPC count. The FNC-PAE also supports efficient procedure call and return, stack operations and branching. Up to three independent jump targets can be evaluated in a single cycle.

The *Side Function Unit (SFU)* operates in parallel to the ALU datapath. It supports 16×16-bit Multiply Accumulate (MAC) functions with 32-bit results and bit-field extraction. The SFU delivers its results directly to the register file.

For efficient code access, a local 1,024 × 256-bit 2-way set-associative L1 instruction cache (I-cache) is provided. Memory accesses are performed by the 32-bit address generator which accesses the tightly-coupled memory (TCM, 4K×16-bit) or external RAM. A Block Move Unit transfers blocks of data between external memory and the TCM in the background. The TCM can also be configured as L1 Data Cache (but this is not implemented in MORPHEUS). Code and data accesses to the external memory hierarchy (e.g. the CEB and external SRAM) utilize dedicated 64-bit wide *Memory Channels*. Furthermore a vectorized Interrupt controller which supports breakpoints and several timers are available within the FNC-PAE.

6.5 XPP-III Components

XPP-III Components provide a flexible and high bandwidth communication framework that links the processing kernels to the outside world. The following components are implemented in the MORPHEUS SoC:

- Crossbars (*XBars*) are used to build a circuit switched network for data streams. XBars can have up to 31 input or output streams and are programmable.
- The *Configuration DMA* requests configuration data from memory (e.g. CEB) and reconfigures the XPP Array.
- The *4D-DMA* controllers provide versatile address generators for complex memory access patterns which are often required for video algorithms. The address generators support clipping and can combine up to four data streams for maximum bandwidth.
- *Memory Arbiters* collect and arbitrate memory requests and distribute them independently to memory outputs. The Memory Arbiter is fully pipelined and supports burst memory accesses and programmable prioritization.
- The *XRAM* is a 8K × 64-bit on-chip memory which is used to buffer data which cannot be stored locally on the XPP Array.

Other components are available off-the-shelf but not implemented in the MORPHEUS chip due to area restrictions. Those components are *Level-2 Caches* between the FNC-PAEs and the Memory Arbiter, *Linear DMA* controllers, an *Interrupt Handler*, *Stream-Fifos*, *Stream-IO* for asynchronous I/O of data streams, and the *RAM-IO* which allows programs running on the XPP Array to directly address external memories.

Standard busses such as AMBA would not be optimal for multi-channel streaming data and do not provide implicit data synchronization. Therefore all XPP-III Components communicate with two types of point-to-point links: *Data Streams* and *Memory Channels*. Data Streams are identical to those which connect the PAEs within the XPP Array. Memory Channels are split into a requester and response part. Requesters have 64/32 bit data/address paths for read or write requests, while the response channel provides 64-bit read data. Both parts are fully pipelined. Thus a number of requests can be issued to memory without the need to wait for memory read responses after every request. A hardware protocol similar to the data stream protocols guarantees full data synchronization. Additionally, the pipelined *FNC-IO-Bus* provides FNC-PAE access to the configuration registers of the XPP III Components.

6.6 XPP-III in the Context of the MORPHEUS SOC

Figure 6.3 depicts the XPP-III HRE in the MORPHEUS SoC. XPP's processing resources are the XPP Array (with 5×6 ALU-PAEs and 2×6 RAM-PAEs) and two FNC-PAEs with local L1 Instruction caches and TCM. All data streams are connected through programmable XBars. The XPP Array configuration is loaded from CEB or external SRAM via the Config-DMA controller. 4D-DMA controllers

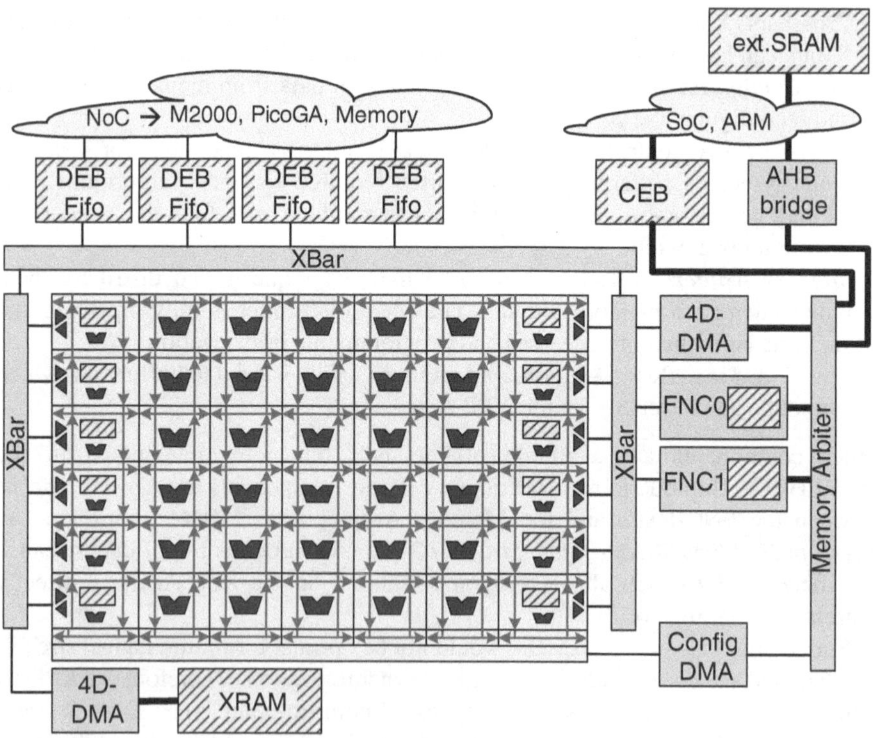

Fig. 6.3 The XPP-III HRE and interfaces to the SoC

convert data streams to memory access patterns for either the local XRAM buffer or the external SRAM or the Configuration Exchange Buffer (CEB).

According to the MORPHEUS concept, four independent data streams are connected to Data Exchange Buffers (DEBs). XPP's data streams fit perfectly to the transfer mechanism of the NoC. Therefore the DEBs can be simple Fifos (one for each direction) and do not need further software-controlled synchronization mechanisms for ping-pong buffer transfer.

The ARM processor loads the XPP Application code into the CEB. The application code is the binary generated by the XPP tool chain. It includes code for both, the FNC-PAEs and the XPP Array. If the CEB size is too small for an application, the code can also be located in external memory which is mapped into the XPP address space. Since applications running on XPP are not limited to process only streaming data originating from the DEBs, XPP can also directly access data within the external SRAM. However, in the MORPHEUS SoC the interface to external memory is not optimized for bandwidth. Additionally, the MORPHEUS Exchange Registers (XRs) are mapped into the SRAM address space.

6.7 Software Development Overview

The XPP-III architecture is fully scalable and can be tailored to the specific needs of an application domain. In the following sections we give an introduction to XPP programming and tools.

The tools support all features of the XPP-III architecture even if not all of them are available in the MORPHEUS hardware. Therefore a special tool version was compiled that takes the restricted MORPHEUS setup into account. Hence, when planning applications for the XPP-III HRE, the programmer must be aware of the MOPRHEUS hardware, such as the size of the XPP Array, the communication channels, the number of FNC-PAEs, the available XPP Components, and the potential restrictions.

Figure 6.4 gives an overview of the typical application development process. Any C application can be directly compiled to a FNC-PAE and run on it. However, in order to achieve the full XPP-III performance, partitioning of the application code into code sections running on one or more FNC-PAEs and on the XPP Array is required.

For a good partitioning, the sequential code is first profiled. Based on the profiling results, the most time-consuming function calls and inner program loops are identified. These code sections are likely candidates for acceleration on the XPP Array, especially if they are *regular*, i.e. if the same computations are performed on many data items. We call them *dataflow sections* since the computations can be performed by data streaming through dataflow graphs. In the C/C++ code, these sections are typically represented as loops with high iteration counts, but with few conditional branches, function calls or pointer accesses. These program parts exhibit a high degree of loop-level parallelism. Note that the ALU- and RAM-PAEs

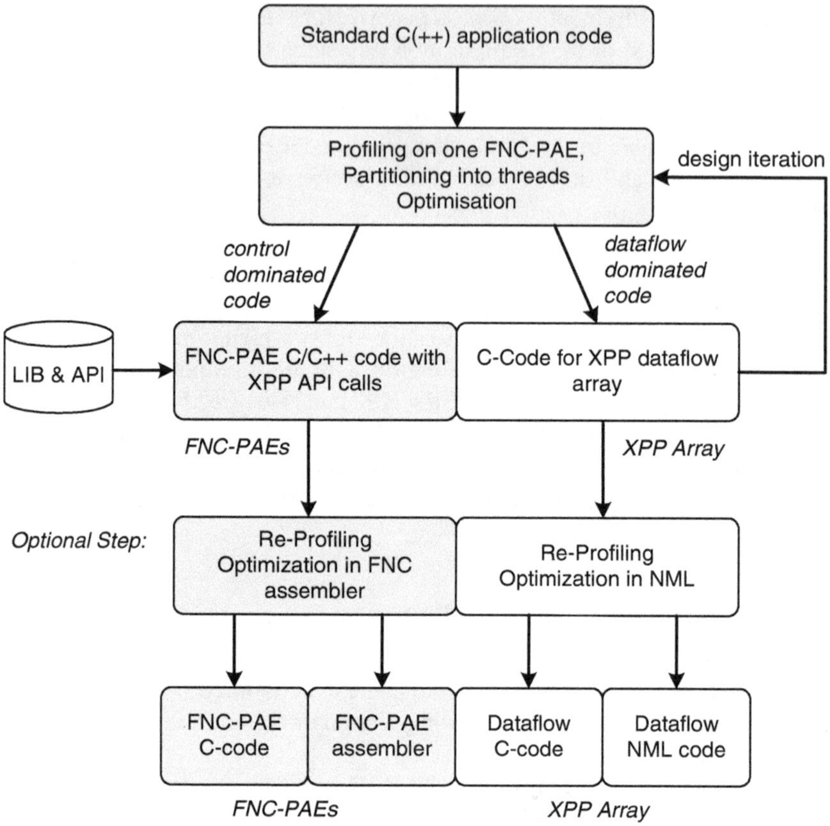

Fig. 6.4 XPP-III software development overview

are not restricted to processing pure dataflow graphs (DFGs). They can handle nested loops and nested conditions as well.

If time-consuming *irregular* code exists in the application, a coarse-grain parallelization into several FNC-PAE threads can also be very useful. This even allows running irregular, control-dominated code in parallel on several FNC-PAEs. Again, XPP API calls are used for communication and synchronization between the threads. Semaphores are also provided to guarantee exclusive access to shared resources like memory or I/O. The threads mapped to FNC-PAEs can be further optimized by using assembler libraries or by writing critical routines directly in FNC-PAE assembler.

6.8 XPP Array Programming

PACT's *XPP Vectorizing C Compiler* (XPP-VC) provides the fastest way to generate XPP Array configurations. It directly translates standard C functions to XPP configurations. The original application code can be reused but may require some adaptations since XPP-VC cannot handle C++ constructs, structs, and floating-point operations. Furthermore, specific XPP I/O functions (corresponding to the XPP API calls on the FNC-PAEs) must be used for synchronization and for data transfers.

The XPP-VC compiler uses vectorization techniques to execute suitable program loops in a pipelined fashion, i. e. data streams taken from RAM-PAEs or from I/O ports flow through operator networks. In this way many ALUs are continuously and concurrently active, exploiting the XPP dataflow array's high performance potential [6].

6.8.1 XPP-VC Code Example

The C code in Fig. 6.5 is a small *for* loop with a conditional assignment and a XPP I/O function for a port output. The XPP functions are defined in file *XPP.h*. The right side shows the dataflow graph for this program generated by XPP-VC. While the counter *COUNT* controls the loop execution, the comparator *LT* and the multiplexer *SWAP* select the result being forwarded to the output port *DOUT0*. The dotted arrow from *LT* to *SWAP* is an event (i.e. 1-bit control) connection. All other connections are data connections.

Especially on relatively small implementations like the MORPHEUS Array, XPP-VC may require more PAEs than available. In this case or if a highly optimized implementation for a dataflow section is required, the code can be directly implemented in NML. The effort is comparable to assembler programming and

```
#include "XPP.h"
#define N 10

main() {
  int i, res;

  for (i = 0; i<N; i++) {
    if (i < 5)
      res = i;
    else
      res = 2 * i - 3;
  XPP_putstream(LEFT, 4, 0, res);
  }
}
```

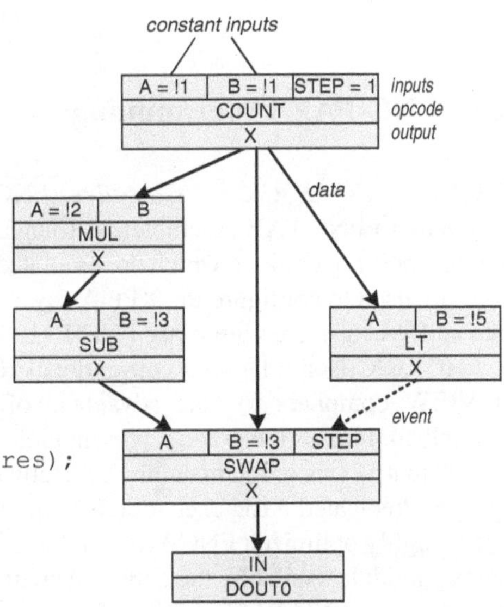

Fig. 6.5 XPP-VC loop example

much simpler than HDL (e.g. Verilog or VHDL) design. In contrast to HDL design, only the functionality of the dataflow section needs to be described in NML, but no timing issues arise.

In NML, arithmetic expressions are described like C expressions and automatically converted to operator trees. Counters, memories, dataflow operators (stream multiplexers, demultiplexers, mergers, etc.), accumulators, and event operators (for program control) are explicitly allocated and connected to other operators or expressions. They are used to implement conditional computations (branches) and iterative computations (loops). In this way, the control flow of a C function can be mapped to the XPP Array as well. Instead of using arithmetic expressions, all operators can also be explicitly allocated. This allows to place them manually on the XPP Array. Hierarchical modules allow component reuse, especially for repetitive layouts.

6.8.2 NML Code Example

The NML code in Fig. 6.6 is a direct NML implementation of the dataflow graph in Fig. 6.5. Here, the else branch of the condition is described as an arithmetic expression, and all other operators are explicitly allocated. Note that the XPP statement at the top of the file defines the XPP core parameters (e.g. for the MORPHEUS hardware), and the *DELAY_BALANCE* command is used for pipeline balancing to optimize the throughput of the loop body. As with XPP-VC, it is possible to instantiate NML Library modules in a manually designed NML configuration to achieve maximum performance with reduced programming effort.

6.9 FNC-PAE Programming

PACT's *FNC-PAE C/C++ Compiler* (FNC-GCC) compiles ANSI C and C++ programs to FNC-PAE assembler. All language features are supported, including floating-point operations which are emulated by the integer ALUs. XPP API functions are used to configure the XPP Array, to communicate with it, or to communicate and synchronize with other FNC-PAE threads.

FNC-GCC is similar to a conventional RISC compiler, but uses some features of VLIW compilers to take advantage of the code's intrinsic instruction-level parallelism. It maps the graph representation of the input program to the FNC-PAE's 8-ALU matrix (via graph matching), thereby utilizing as many ALUs as possible per opcode. Predicated ALU execution is employed for small if-then-else constructs.

If a highly optimized FNC-PAE implementation is required, but no suitable FNC Library module available, the code can be directly implemented in FNC assembler. It supports all FNC-PAE hardware features.

The assembler uses three-address code for most instructions, as in this example: *SUB target, source1, source 2*.

```
XPP(V3.3, 7,6, 8,6;
  DATA_BIT_WIDTH = 16 CONFIG_BIT_WIDTH = 24
  FREG_DATA_PORTS = 4 BREG_DATA_PORTS = 4
  FREG_EVENT_PORTS = 4 BREG_EVENT_PORTS = 4
  IRAM_ADR_WIDTH = 9)// Morpheus Array Setup

MODULE forexample {
    OBJ ctr: COUNT { // loop counter
    A =! 1
    B =! 9
    STEP = 1
    }
  OBJ cond: LT { // loop condition
    A = ctr.X
    B =! 5
    }

   // arithmetic expression for else branch
  SIG DATA elsebranch// signal definition
  elsebranch= EXPR(2 * ctr.X - 3)

  OBJ mux: SWAP { // result multiplexer
    A = elsebranch
    B = ctr.X
    STEP = cond.U
    }
  OBJ outport: DOUT0 { // output port
    IN = mux.X
    }

    // pipeline delay balancing command
    DELAY_BALANCE(ctr -> outport)
}
```

Fig. 6.6 NML loop example

Multiple ALU instructions are merged into one FNC opcode as follows: The instructions for the left and right ALU columns are separated by a vertical bar (|), and the ALU rows (at most four) are just described one by one. A FNC opcode is terminated by the keyword *NEXT*.

6.9.1 FNC-Assembler Example

The FNC assembler code in Fig. 6.7 sequentially multiplies two 8-bit numbers (in registers *r0* and *r1*, with the 16-bit result in *r2*. Note that this example was only chosen for demonstration purposes. In a real application, the single-cycle 16×16 MUL instruction in the SFU would be used instead.

The first opcode initializes the registers, including the loop counter *r7*. The second opcode (after the label *loop*) contains all loop computations, including counter

```
; initialize parameters for test
      MOV r0, #10 ; operand 0
      MOV r1, #6  ; operand 1
      MOV r2, #0  ; clear result register
      MOV r7, #8  ; loop counter init
      NEXT

loop:
      SHRU r0, r0, #1    | SHL r1, r1, #1
 CY   ADD  r2, r2, r1
 ACT  SUB  r7, r7, #1
 ZE   NOP  ! HPC loop
      NEXT
...
```

Fig. 6.7 FNC assembler example

decrement, test and jump. The predicates before the instructions have the following meanings: *CY* indicates that *ADD* is only executed if the shift *SHRU* above it had a carry-out value one, and *ACT* means that *SUB* is executed (activated) in any case. *ZE NOP ! HPC loop* instructs the FNC-PAE to perform a single-cycle jump to label *loop* (*high-performance continue* = HPC, " ! " reads as "else") if the *SUB* instruction above it did not set the zero flag (*ZE*). This means that every loop iteration requires only one cycle. If *r7* is zero, i.e. the *ZE* flag set, the program execution continues after the loop.

6.10 XPP Software Development Tools (XPP-III SDK)

The XPP-III tool chain provides all features for code entry, simulation and execution on the MORPHEUS hardware. Figure 6.8 shows the *XPP-III-SDK* tool flow starting with the partitioned C-code. *XPP-VC* compiles C to NML code, and *FNC-GCC* compiles C/C++ to FNC assembler. All FNC assembler files are processed by the FNC-PAE assembler *XFNCASM*. NML files (whether generated by *XPP-VC*, manually designed, or from the NML Module Library) are processed by the XPP mapper *XMAP*. It compiles NML source files, automatically places and routes the configurations, and generates XBIN binary files.

Finally, the FNC and XBIN binaries and the XPP API Library are linked to an XPP application binary. It is executed and debugged with the cycle accurate XPP-III SystemC simulator *XSIM*. On the MORPHEUS hardware, the ARM processor loads this binary either to the CEB or a reserved memory area in external SRAM and starts the XPP by issuing a command to the exchange registers according to the MORPHEUS concept. Note that the XPP Simulator implements the DEBs as File I/O.

In both cases, the application can be visualized and debugged by the XPP debugger. This tool visualizes the data being processed on the XPP Array and the FNC-PAEs cycle by cycle. The debug communication with the MORPHEUS SoC is performed by a dedicated JTAG port for the XPP HRE.

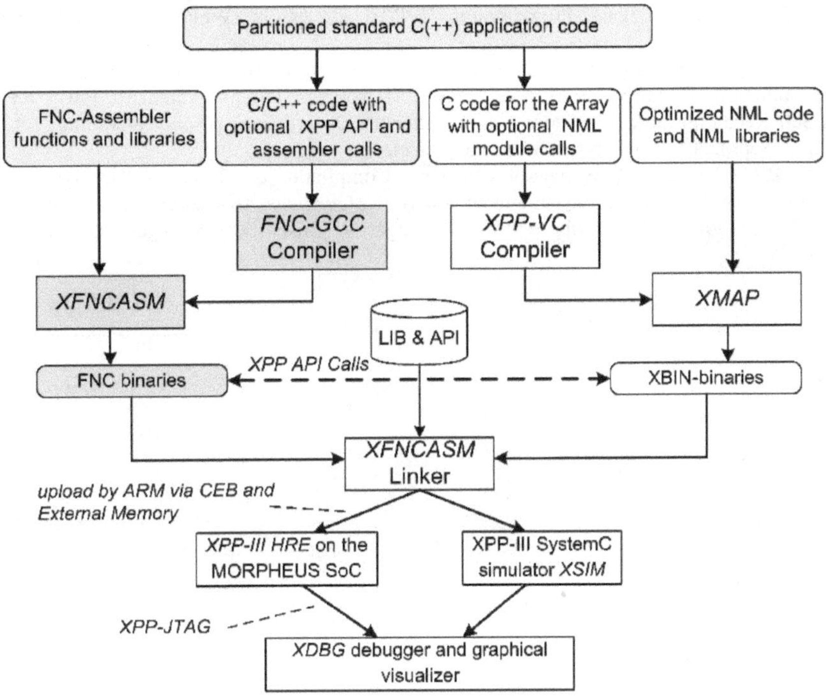

Fig. 6.8 The XPP-III tool chain

6.11 Conclusions

The XPP-III reconfigurable HRE integrated into the MORPHEUS SoC provides the flexibility and performance required for the wide range of applications targeted by this SoC. In the MORPHEUS project it is mainly used as a high-bandwidth accelerator for streaming data originating from the NoC and other HREs. Since XPP-III may also be used as a standalone processor core without relying on the control services of the ARM processor, it can not only execute single accelerator functions but also complete applications (e.g. video decoders). Since the MORPHEUS SoC is only a technology demonstrator, not all features of the XPP-III IP have been implemented. Nevertheless, the XPP-III SDK design tools allow simulating and evaluating more complex and larger architectural XPP-III designs as well.

References

1. PACT XPP Technologies, XPP-III Processor Overview (White Paper), 2006, www.pactxpp. com.
2. PACT XPP Technologies, Reconfiguration on XPP-III Processors (White Paper), 2006 www. pactxpp.com.

3. PACT XPP Technologies, Programming XPP-III Processors (White Paper), 2006 www.pactxpp.com.
4. PACT XPP Technologies, Video Decoding on XPP-III (White Paper), 2006 www.pactxpp.com.
5. V. Baumgarte, G. Ehlers, F. May, A. Nückel, M. Vorbach, and M. Weinhardt, PACT XPP – A Self-Reconfigurable Data Processing Architecture, *The Journal of Supercomputing*, Vol. 26, No. 2, Sept. 2003, Kluwer Academic Publishers.
6. J. M. P. Cardoso and M. Weinhardt, Chapter 9, Compilation and Temporal Partitioning for a Coarse-Grain Reconfigurable Architecture, in *New Algorithms, Architectures and Applications for Reconfigurable Computing* (editors: P. Lysaght, W. Rosenstiel), Springer, Dordrecht, NL, 2005.

Chapter 7
The Hardware Services

Stéphane Guyetant, Stéphane Chevobbe, Sean Whitty, Henning Sahlbach, and Rolf Ernst

Abstract High-end applications have been designed for the MORPHEUS computing platform to fully demonstrate its potential as a high-performance reconfigurable architecture. These applications are characterized by demanding memory bandwidth requirements, as well as multiple processing stages that necessitate dynamic reconfiguration of the heterogeneous processing engines. Two hardware services have been specifically designed to meet these requirements. This Chapter first describes the unit responsible for reconfiguration of the various processing engines presented in Chapters 4–6 and the predictive method used to hide reconfiguration latencies. The second part of this Chapter describes a bandwidth-optimized DDR-SDRAM memory controller, which has been designed for the MORPHEUS platform and its Network On Chip interconnect in order to meet massive memory throughput requirements and to eliminate external memory bottlenecks.

Keywords Bandwidth • bank interleaving • caching • CMC • configuration overhead • external memory • DDR • HW task allocation • latency • memory access • memory controller • predictive prefetch • QoS • Quality of Service • reconfiguration manager • request bundling • SDRAM • throughput

7.1 Predictive Configuration Manager

A new class of dynamically reconfigurable multi-core architectures has emerged since a decade, able to cope with changing requirements. The dynamic reconfiguration allows changing the hardware configuration during the execution of the tasks.

S. Guyetant (✉) and S. Chevobbe
CEA LIST, Embedded Computing Laboratory, France
stephane.guyetant@cea.fr

S. Whitty (✉), H. Sahlbach, and R. Ernst
IDA, TU Braunschweig, Germany
whitty@ida.ing.tu-bs.de

N. Voros et al. (eds.), *Dynamic System Reconfiguration in Heterogeneous Platforms*, Lecture Notes in Electrical Engineering 40,

This attractive idea of time-multiplexing in reconfigurable architecture does not come for free. Two main approaches address the challenge of dynamic reconfiguration, one is known as temporal partitioning (the system specification must be partitioned into temporal exclusive segments called reconfiguration context), an other one is to find an execution order for a set of tasks that meets system design objectives (known as dynamic reconfigurable logic multi-context scheduling).

With devices that have the capability of run time reconfiguration (RTR), multitasking is possible and very high silicon reusability can be achieved. This can significantly improve the computing efficiency, but RTR may result in configuration overhead, in terms of latency and power consumption, which can largely degrade the overall performance.

7.1.1 Related Work

7.1.1.1 Motivations for Dynamic Scheduling

The main reason that justifies the interest of run time scheduling techniques is that the behaviour of some applications is non deterministic. A growing class of embedded systems needs to execute multiple applications concurrently with high dynamic behaviour (i.e. created by user and dependent on data) [1].

Without *a priori* knowledge of future application workloads, an OS must make decisions based on incomplete information [2]. Systems using WCET (worst case execution times) estimates to perform a static scheduling could be highly underutilized. A compromise is with hybrid solutions that reduce the runtime computations while providing high quality schedules.

7.1.1.2 Configuration Prefetch and Caching

The configuration prefetch consist in retrieving the large configuration data in advance into a configuration memory or directly inside an unused part of the reconfigurable area, before it is actually needed for execution, so that the transfer time is hidden, at least partially.

A way to provide high-quality schedules with a small runtime overhead, [3] proposes a mixed approach with both static and run-time analysis simplified thanks to a heuristic.

At design time each node of each subtask graph is tagged with a weight that represents how critical that node's execution is. The module computes weights by performing an as-late-as possible scheduling. The designer analyzes subtask graphs at design time, and can force critical nodes.

The process of runtime scheduling starts with a schedule that neglects the reconfiguration latency. The reconfiguration manager then updates the schedule by including the necessary reconfiguration times and minimizes their latency overheads.

The configuration caching is similar to the instruction caching found in processors: the goal is still to have the reused control information present in on chip memories when needed, and thus, the cost of loading bitstreams on a reconfigurable core is lower. Several algorithms for configuration caching are developed in [4], targeting various models of FPGAs: in particular, a multi-context FPGA and partially reconfigurable FPGAs are studied.

Reference [5] extends the cache locking technique to configuration locking. The basic idea is to track at run time the number of times that tasks are executed and always lock a number of the most frequently used tasks on dynamically reconfigurable hardware to prevent them from being evicted by the less frequently used tasks. To avoid the overlocking behaviour, they developed an algorithm that can be used at run-time to estimate the configuration lock ratio.

7.1.1.3 FPGA Specific Techniques

A particular class of reconfigurable SOCs is based on the dynamic partial FPGA reconfiguration technology, which supports both 1D reconfiguration, where each task occupies a contiguous set of columns, and 2D reconfiguration, where each task occupies a rectangular area. Real-time scheduling for 1D reconfigurable FPGAs shares many similarities with global scheduling on identical multiprocessors. But hardware task scheduling on FPGA is a more general and difficult problem than multiprocessor scheduling, since each hardware task may occupy a different area size on the FPGA, and even be implemented in several partial bitmaps with different shapes. A task scheduler and placer are needed to find empty space to place a new task, and recycle the occupied area when a task is finished while making sure all task deadlines are met.

Fine grain architectures are particularly penalized, because their reconfiguration carries a significant overhead in the range of milliseconds that is proportional to size of the area being reconfigured. [6] proposes particular partitioning and scheduling algorithms integrating constraints such as: task must fit in FPGA area; source node in task graph does not need the reconfiguration stage; a task can only start its execution stage after its reconfiguration stage; two tasks can overlap with each other on either the vertical time axis of the horizontal axis but not both; reconfiguration controller is a shared resource, so reconfiguration stages of different tasks must be serialized; an execution starts only after the end of the previous one; tasks must finish before schedule length.

7.1.2 Presentation of the PCM

The dynamic reconfiguration mechanism described in this section was designed according to the specifications of a heterogeneous architecture such as the MORPHEUS platform. The PCM basically hides the context switching overhead due to the configuration time of a new task on the Heterogeneous Reconfigurable

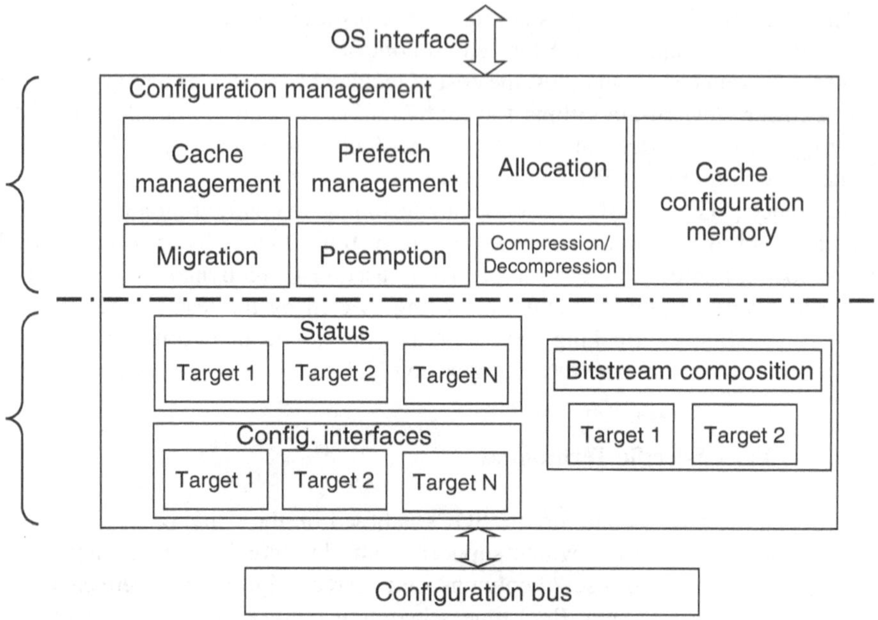

Fig. 7.1 Diagram of an ideal configuration manager

Engine by implementing prefetch and configuration caching services. An ideal view of the PCM is shown in Fig. 7.1. It makes appear a clear separation between the system level and the hardware level. One of the key point of the PCM is to abstract the interface of the reconfigurable engine from the designer point of view.

This method is based on an intermediate graphical representation of the applications extracted at design time, where the nodes represent the functions that are mapped on the reconfigurable engines. Note that the PCM does not handle the instruction-level reconfiguration inside the reconfigurable engines for those that have this property: more precisely, the cycle reconfiguration of the DREAM processor or the partial reconfiguration of the XPP array, are not targeted by the PCM service. On the contrary, the function-level dynamic reconfiguration described hereafter operates at a higher time-frame granularity that allows more complex reconfiguration procedure than a DMA-like fetch or autonomous caching. This service is based on a hybrid design time/run time solution, based on characteristics extracted at design time and run time information given by the scheduler. The prediction method of the PCM is computed during the execution time of the applicative tasks, so that even wrong predictions do not impact the overall execution time.

7.1.2.1 Static Extraction of Core-Dependent Configuration Graphs

The Molen compilation flow provides configuration call graphs (CCG) at the thread level, derived from pragmas in the applicative source code. These pragmas are inserted by the application designer in order to explicit the functions that will be accelerated on the reconfigurable cores. The resulting graphs for all threads of the application are provided to the operating system that computes the run time schedule of the tasks. This procedure will be explained in Chapter 12 and is therefore not detailed here: we focus on how the PCM can leverage the operating system of the configuration task and helps to alleviate the reconfiguration overhead.

The CCGs issued at compile time are *uncoloured*, that is to say, they do not contain the information of the target reconfigurable engine on which the task can be executed. This property gives the designer the flexibility to create different implementations for the accelerated functions, not only for various engines, but also several implementations for the same engine, for example a performance driven implementation with a high level of parallelism, and an other one driven by area constraints, running slower the same task with an iterative process. Even software implementations can be provided if spare processors or DSP cores are included in the heterogeneous platform. The allocation of the task on the hardware is predicted dynamically by the PCM that has the knowledge of the pressure on each computing node.

For each kind of computing engine present in the platform, a core-dependent, or *coloured*, configuration call graph is extracted from the CCG graph, by keeping only the nodes that correspond to an existing implementation on the considered core. The created edges are associated with a weight that corresponds to the number of nodes that have been simplified from the whole CCG. The profiling static information contained in the CCG is kept and enhanced with an implementation priority given by the designer of the reconfigurable implementation. Figure 7.2 illustrates the generation of such graph.

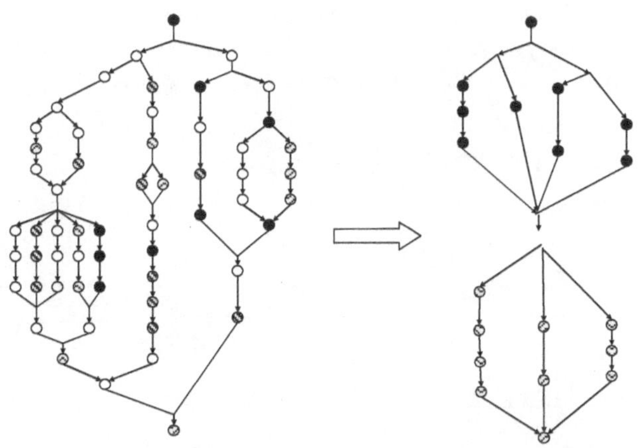

Fig. 7.2 Extraction of core-dependent graphs from a CCG; this example shows the extraction of two sub graphs out the four possible

7.1.2.2 The PCM Interfaces

The PCM is a high level service offered to the OS: it receives configuration control commands, namely the prefetch command, to indicate that a configuration is queued, the execution command when the operation must start, and the release command when a configuration will not be used in the near term.

In addition to the configuration control commands, the PCM service interacts with the scheduler thanks to a set of requests issued by the scheduler. Typically, these requests provide the scheduler with status information of the memory contents, reflecting the prefetched configurations. More advanced requests are computed by the PCM: for example, the "*TimeToExecute*" request returns the remaining time needed before a configuration is ready to be executed, estimated from the respective size of the partial bitstreams still in the configuration memory hierarchy; this time is bounded by zero if the bitstream was already prefetched inside the reconfigurable core, to the maximal time to copy the entire bitstream from the external memory.

As described in Fig. 7.3, the PCM services receives dynamic information from the OS, that are mainly the configuration commands, but also the thread priorities that are used by the prefetch service. The static information, such as execution probability (extracted from application profiling and that annotate conditional branches) and implementation priority (given by the application designer to differentiate several implementations of the same software function for example on heterogeneous cores), are embedded in the graph representation; so that they are easily retrieved by the PCM. The allocation, caching and prefetch services are then ultimately translated into commands to transfer the bitstreams between the levels of the configuration memory hierarchy.

The configuration service is meant to deal with every reconfigurable core, and not only those selected for the MORPHEUS implementation; this explains why it does not provide specialized decompression service nor is it intrusive with the internal configuration mechanisms: the goal is to provide a unified access to the configuration interfaces at the system level. All existing reconfigurable cores have their own protocol; anyway, they can be classified in two main categories. The first include the loaders that are passive (memory-mapped or decoding frames). Active

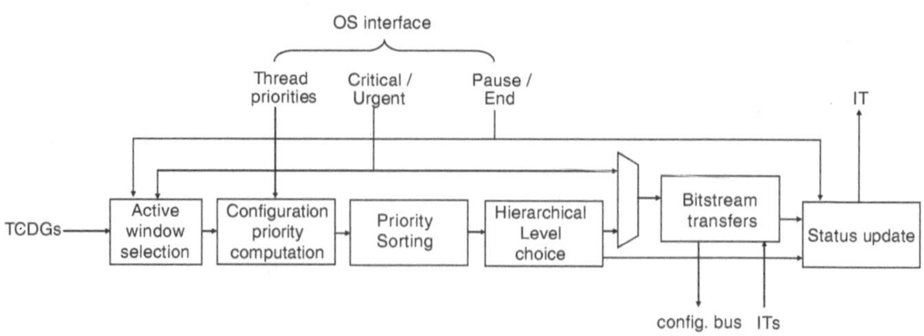

Fig. 7.3 Functional structure of the Predictive Configuration Manager

loaders, that are autonomous in retrieving their configuration data, belong to the second category.

For the MORPHEUS platform, the PCM service is able to prefetch configurations internally to the passive loaders, but restricts the prefetch for active loaders at the cache memory level.

7.1.2.3 Predictive Reconfiguration

The predictive chain is composed of two parts. One is responsible of the selection of the next tasks that are candidates for the next transfers; the other one is responsible of configuration memory management as shown in Fig. 7.3.

In an autonomous way from the scheduler, the PCM does not only selects the next tasks that are to be queued for scheduling, but does a broader search inside all configuration call graphs: let us consider the example presented in Fig. 7.2; the first task is black: at the end of this task, the "black" core is released, but the next black tasks are too deep in the original graph to be selected for immediate prefetch. Instead, the PCM is able to walk the core-dependant graph looking for the next tasks that have an existing implementation for the just freed core. This function is referred as the "Active Window Selection" in Fig. 7.3. It analyses the applications graphs to determine a first set of the next candidates.

Then, they are affected by a dynamic priority calculated from the static and dynamic parameters in the function "Configuration priority computation". This done by a polynomial function implemented with coefficients that can be tuned by the designer according to different application behaviours. In a third step, these dynamic priorities are sorted together (by the function "Priority sorting") so that the most relevant prefetch actions can be sorted inside a FIFO.

In the last step, following this order, the bitstream transfers can start until a new command is issued by the scheduler. To maintain an accurate view of the memory state the PCM update its status memory registers after each transfer. Obviously, if the execution time between two consecutive schedules is too short, the prefetch can not take place, but at least the service does not create additional overhead.

An important feature of the PCM is that the prediction method always resynchronizes with the actual schedule provided by the operating system. As the PCM behaves as a slave of the operating system, its prefetch can always be overridden, at the cost of a misprediction: this is the case of a false negative prediction, and the reconfiguration time might not be masked. Then the next prediction is recalculated from the new status and does not stay into a wrong prediction branch.

Another robustness feature was implemented to ensure that false positive predictions are removed from the memory hierarchy. In the nominal case, configuration blocks are only deleted to be replaced by newer predictions, but all blocks present in the prediction list are protected by a flag, and this protection is removed after a release command, or when an OR divergence was chosen in the configuration graphs. In the same time, configurations are associated with an age that represents the time from which they are present inside the on chip hierarchy. This age is reset at each time that an execution command is issued by the operating system.

The block replacement scheme then deleted the oldest unprotected blocks, but also the unused protected blocks present since a parameterized threshold. Finally, the whole behaviour of the PCM can be tuned by a set of parameters to adapt its prefetch and configuration policies according to the profile of the applications. These parameters are set at boot time and can not be change during the execution.

7.1.3 Conclusions

A predictive reconfiguration management service has been presented in this chapter; compared to the related works in the literature, it has been designed to fulfil the particular needs of the MORPHEUS heterogeneous platform composed of heterogeneous reconfigurable cores. As this platform can be scaled to explore the domain of multi-core SOCs based on reconfigurable or programmable nodes, the PCM can handle other flavours of computing engines, and several kinds of each core. As it handles the complexity of the heterogeneous allocation, the scalability of the platform is not a burden for the OS.

The power consumption of the PCM was not studied yet. As for every caching or prediction mechanism, the miss rate is an important value, especially because the bitstreams involved have a size in the order of dozens of kilobytes. Future enhancement of the PCM should contain policies that focus on maximizing the reuse rate of prefetched bitstreams and possibly manage the voltage and frequency of the memories involved in the configuration hierarchy. Also the current version of the PCM does not check itself; a self-monitor service could be inserted, able to change the prefetch policies depending on prediction miss rate and memory pollution metrics.

7.2 Custom DDR-SDRAM Memory Controller

7.2.1 Introduction

The potential of the MORPHEUS platform will be demonstrated in several application domains. These include reconfigurable broadband wireless access and network routing systems, processing for intelligent cameras used in security applications, and film grain noise reduction for use in high definition video. Unlike some applications, the image-based applications have been shown to exhibit immense memory needs. For example, for real-time operation, digital film applications using the current standard 2K[1] resolution require read data rates of at least 2.3 Gbit/s to load

[1] 2K implies 2048 × 1568 pixels/frame at 30 bits/pixel, and 24 frames/s for real-time processing.

a single frame for processing, and write data writes of at least 2.3 Gbit/s to write the processed image back to memory. This number can significantly increase when frequent intermediate frame storage is necessary. Higher resolutions of up to 4K and even 8K are on the horizon, which will also increase data rates.

Satisfying such memory requirements is no easy task, and SDRAM interfaces have long been a critical performance bottleneck [7]. However, by taking advantage of memory access optimizations, these limitations can be greatly reduced. In the MORPHEUS project, a bandwidth-optimized custom DDR-SDRAM memory controller was designed in order to meet the external memory requirements of each of the planned applications.

7.2.2 Requirements

The MORPHEUS project implements a key step in the digital film processing chain: film grain noise reduction. This application has been previously implemented in the FlexFilm project [8]. In the MORPHEUS project, the noise reduction application will be mapped across all three heterogeneous reconfigurable entities, and must process at least 3 MPixel/frame and 24 frames/s (assuming a 2,048 × 1,556 resolution). This results in approximately 5 to 170 GOPS, depending on the complexity of the chosen algorithm.

Additionally, as the application input consists of streaming image data, the algorithms require a significant amount of memory, specifically for the frame buffers required by the motion estimation/motion compensation stage and the synchronization buffers needed by the discrete wavelet transform filtering. These buffers are too large for internal RAM. Consequently, a large external SDRAM is necessary to meet storage requirements. Existing external memory controller solutions can support large SDRAMs, but not the high throughput requirements demanded by such applications. Therefore, a custom design was created for the MORPHEUS architecture.

7.2.3 Architecture

The MORPHEUS DDR-SDRAM controller (CMC)[2] consists of three main components: the NoC-CMC interface, the Two-stage Buffered Memory Access Scheduler, and the DDR-SDRAM interface. An architectural overview is shown in Fig. 7.4. The overall architecture and the controller core, also known as the Two-stage Buffered Memory Access Scheduler, are described in the following sections. For a more detailed description of the memory controller architecture, including details of the NoC-CMC and DDR-SDRAM interfaces, consult [9,10].

[2] CMC stands for Central Memory Controller.

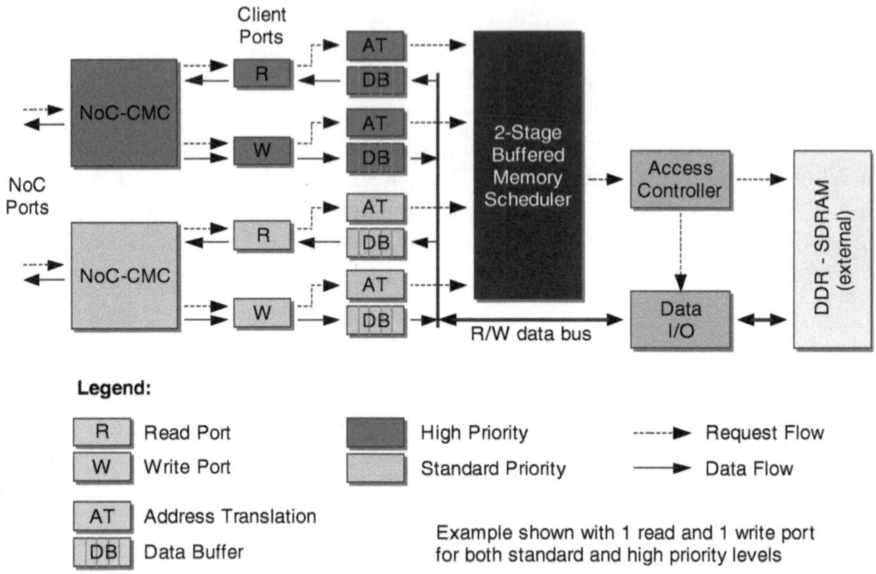

Fig. 7.4 SDRAM controller architecture

7.2.3.1 General Architecture

Memory access requests to the SDRAM controller are made by applications via the MORPHEUS Data Protocol interface, which provides the connection from the MOPRHEUS Network On Chip (NoC) to a configurable number (up to 8) of application read and write ports. The MORPHEUS NoC is based on the STNoC Network On Chip described in [11].

Many applications can perform concurrent memory accesses; however, it is *not guaranteed* that requests *to the same memory address from different ports* are executed in order (see Section 7.2.3.4).

Memory requests first enter the NoC-CMC Interface, where read and write requests from MORPHEUS applications are buffered and converted from NoC packets to regular CMC read and write requests and sent to the CMC in burst request format. A CMC burst consists of 8 consecutive data words, while a data word is 64 bits in length.

After entering the CMC, memory access requests first reach the Address Translator, where the logical address is translated into the physical bank/row/column quadruple required by the SDRAM. Concurrently, at the Data Buffers, the write request data is stored until the request has been scheduled; for read requests a buffer slot for the data read from the SDRAM is reserved.

The requests then enter the core part of the SDRAM controller, the Two-stage Buffered Memory Access Scheduler (see Section 7.2.3.2). After one request is selected, it is executed by the Access Controller and the data transfer to/from the corresponding data buffer is initiated by the Data I/O module. Finally, external data transport and signal synchronization for DDR transfers is managed by the DDR Interface and its 64-bit data bus.

7.2.3.2 Two-Stage Buffered Memory Access Scheduler

The Two-stage Buffered Memory Access Scheduler comprises the core of the memory controller, performing access optimizations and eventually issuing requests to SDRAM. Figure 7.5 illustrates the scheduling stages.

The single-slot request buffers are used to decouple the clients from the following scheduling stages and can accept one incoming request per clock cycle.

The first scheduler stage, the request scheduler, selects requests from these buffers, one request per two clock cycles, and forwards them to the bank buffer FIFOs. By applying a round-robin arbitration policy, a minimum access service level is guaranteed. As stated above, high priority requests are serviced before standard priority requests when priority levels are enabled.

The bank buffer FIFOs, one for each bank, store the requests according to the addressed bank.

The second scheduler stage, the bank scheduler, selects requests from the bank FIFOs and forwards them to the access controller for execution. In order to increase throughput utilization, the bank scheduler performs *bank interleaving* to hide bank access latencies and *request bundling* to minimize stalls caused by read-write switches.

Bank Interleaving exploits the SDRAM structure, which is organized into independent memory banks. SDRAM banks require 4 (read) to 6 (write) passive cycles after a data transfer, during which the active bank cannot be accessed. By reordering memory requests to ensure consecutive accesses occur to inactive banks, a second bank can be accessed during such idle times, effectively hiding these latencies and significantly increasing data rates.

Request Bundling minimizes the effects of idle cycles required during bus direction switches. These stalls (1 cycle for a read-write change, 1–2 cycles for a write-read

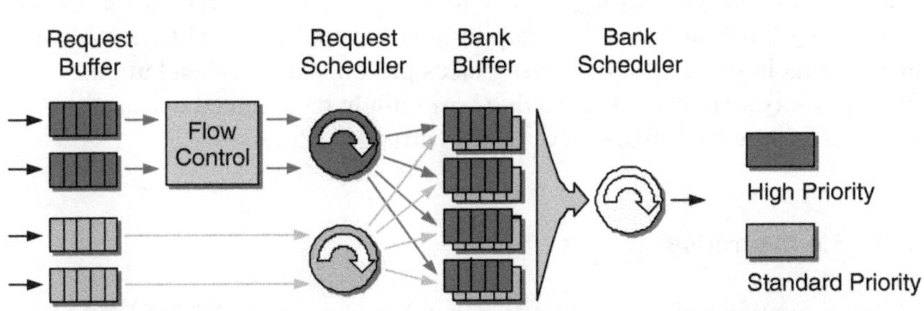

Fig. 7.5 Two-Stage Buffered Memory Access Scheduler

change, depending on the SDRAM module) can decrease overall throughput by up to 27% [12]. By bundling like requests together into continuous blocks, these stalls can be avoided.

7.2.3.3 Quality of Service (QoS)

While not a consideration for the MORPHEUS platform, Quality of Service is important for modern SDRAM controllers. In general, CPU cache miss and data path memory requests show different memory access patterns. For effective operation, CPU cache miss memory accesses should be served with a *smallest possible latency*, while data path memory requests should be served with a *guaranteed minimum throughput at guaranteed maximum latency*. A more detailed explanation can be found in [7].

To handle these requirements, two priority levels for memory access requests have been implemented in the CMC. High priority requests (*smallest possible latency*) are always executed before standard priority requests. This is implemented via distinct access paths for high and standard priority requests and a modified bank scheduler, which always executes high priority requests first.

With any priority-based design, starvation at the lower levels is a potential issue. To avoid starvation of standard priority requests (*guaranteed minimum throughput at guaranteed maximum latency*), a flow control unit is used to reduce the maximum throughput of high priority requests. The flow control unit can be configured to pass n requests within T clock cycles to allow bursty CPU memory accesses when necessary.

7.2.3.4 Memory Coherency

Despite potential reordering of memory access requests during the scheduling stages, steps have been taken to ensure memory coherency.

Reads and writes from *different ports* to the *same addresses* are potentially executed out-of-order. Within the same priority levels and provided that the bank buffers do not fill up, a distance of $2n$ clock cycles, with n being the number of ports per priority level, is sufficient to exclude hazards.

Reads from *one port* to *different addresses* might be executed out-of-order; however, they finish in-order. This implies that the application always receives the requested data in-order. The reordering takes place inside the data buffers.

Writes from *one port* to *different addresses* might be executed out-of-order. This is a non-issue, however, since they occur at different addresses.

7.2.3.5 Configuration

CMC configuration parameters clearly depend on the type of DDR-SDRAM used, the system clock frequency, and overall board layout. For the MORPHEUS CMC, many parameters, such as address bus width, data bus width, and the number of application ports must be determined before logic synthesis. However, a certain degree of flexibility must remain in the MORPHEUS CMC so that it may support different DDR-SDRAM modules and to achieve proper timing under real PCB conditions. To achieve this goal, a programmable Configspace module was created, which allows run-time, user-adjustable configuration of SDRAM timing, SDRAM layout, and of the DDR path delay elements used to generate necessary proper timing behavior for the DDR Interface.

The values selected for the current version of the MORPHEUS chip are displayed in Table 7.1.

The MORPHEUS platform is a complex design with numerous integrated IPs, many of which consume significant chip area resources. The CMC was therefore designed to occupy a relatively small chip area and logic was minimized whenever possible. Based on the configuration shown in Table 7.1, the MORPHEUS CMC resource usage is presented in Table 7.2.

7.2.4 Performance

Using access patterns similar to the streaming patterns generated by the film grain noise reduction algorithm outlined in Chapter 14, both read and write throughput were tested. The MORPHEUS CMC data rates come satisfyingly close to the theoretical maximum DDR throughput values, with a total bandwidth utilization of up to 75%.

Table 7.1 MORPHEUS CMC parameter list

Parameter	Value
Data bus width	64 bit
Word size	64 bit
Burst length	8 words
NoC-CMC client ports	3
Standard priority application ports	6 (3 read, 3 write)
SDRAM address bus width	13 bit (13 row, 10 column)
SDRAM banks	4
Chip selects	2
QoS support	Disabled

Table 7.2 MORPHEUS CMC resource usage

Module	Size (KiloGates)
Cmc_core	46.2
Configspace	1.3
Noc-cmc_port_0	26.2
Noc-cmc_port_1	26.3
Noc-cmc_port_2	26.3
Total synthesizable area	126

Despite the CMC's focus on optimizing throughput, latency should not be ignored. Large buffer depths have a negative effect on latency, as well as the access optimization techniques employed by the schedulers. However, the CMC's internal FIFOs were kept at reasonable sizes to minimize their effect.

The same access patterns used in the throughput experiments were also used to test latency. Because of its burst-oriented design, latencies are identical for write operations of all sizes.

More interesting, however, are read access latencies, which correspond to the time an application must wait for requested data. Read latencies proved to be fully dependent on the size of the read command issued to the controller. As expected, the more data requested, the longer the latency.

A more detailed performance analysis, including comprehensive throughput and latency results, can be found in [9].

7.2.5 Conclusions

In this chapter, a novel bandwidth-optimized SDRAM controller for the MORPHEUS heterogeneous reconfigurable platform has been presented. Through access optimizations and a sophisticated memory access scheduler, it supports applications that demonstrate requirements not met by off-the-shelf memory controllers considered by the project. Most importantly, through achievement of up to 75% of the theoretical maximum DDR data rate, the MORPHEUS CMC can supply the data rates necessary for real-time image processing at 2K resolutions. The research-based evaluation of the MORPHEUS platform does not use the CMC as the external memory controller, but rather a less powerful yet silicon-proven ARM PL175 PrimeCell MultiPort Memory Controller due to pin restrictions and manufacturing costs. This evaluation chip represents a single instantiation of the architecture, which can easily be expanded to include the CMC for future incarnations of the chip.

References

1. Noguera, J. and Badia, R.M., Dynamic run-time HW/SW scheduling techniques for reconfigurable architectures, *Proceedings of the Tenth International Symposium on Hardware/Software Codesign*, Estes Park, Colorado: ACM, 2002, pp. 205–210.
2. Huang, C. and Vahid, F., Dynamic coprocessor management for FPGA-enhanced compute platforms, *Proceedings of the 2008 International Conference on Compilers, Architectures and Synthesis for Embedded Systems*, Atlanta, GA, USA: ACM, 2008, pp. 71–78.
3. Resano, J., et al., Efficiently scheduling runtime reconfigurations, *ACM Transactions on Design Automation of Electronic Systems*, vol. 13, 2008, pp. 1–12.
4. Li, Z., Compton, K., and Hauck, S., Configuration caching management techniques for reconfigurable computing. In *FCCM '00: Proceedings of the 2000 IEEE Symposium on Field-Programmable Custom Computing Machines*, 2000.

5. Qu, Y., Soininen, J., and Nurmi, J., Improving the efficiency of run time reconfigurable devices by configuration locking, *Proceedings of the Conference on Design, Automation and Test in Europe*, Munich, Germany: ACM, 2008, pp. 264–267.
6. Yuan, M., He, X., and Gu, Z., Hardware/software partitioning and static task scheduling on runtime reconfigurable FPGAs using a SMT solver, *Proceedings of the 2008 IEEE Real-Time and Embedded Technology and Applications Symposium – Volume 00*, IEEE Computer Society, 2008, pp. 295–304.
7. Heithecker, S., and Ernst, R., Traffic shaping for an FPGA-based SDRAM controller with complex QoS requirements, *Proceedings of the 43rd Annual Design Automation Conference (DAC)*, 2005, pp. 575–578.
8. do Carmo Lucas, A., Heithecker, S., and Ernst, R., FlexWAFE – A high-end real-time stream processing library for FPGAs, *Proceedings of the 44th Annual Design Automation Conference (DAC)*, 2007, pp. 916–921.
9. Whitty, S., and Ernst, R., A bandwidth optimized SDRAM controller for the MORPHEUS reconfigurable architecture, *Proceedings of the IEEE Parallel and Distributed Processing Symposium (IPDPS)*, 2008.
10. do Carmo Lucas, A., Sahlbach, H., Whitty, S., Heithecker, S. and Ernst, R., Application development with the FlexWAFE real-time stream processing architecture for FPGAs, *ACM Transactions on Embedded Computing Systems, Special Issue on Configuring Algorithms, Processes and Architecture (CAPA)*, 2009.
11. Coppola, M., Locatelli, R., Maruccia, G., Pieralisi, L., and Scandurra, A., Spidergon: a novel on-chip communication network, *Proceedings of the International Symposium on System-on-Chip*, 2004, pp. 16–18.
12. Heithecker, S., do Carmo Lucas, A., and Ernst, R., A mixed QoS SDRAM controller for FPGA-based high-end image processing, *Workshop on Signal Processing Systems Design and Implementation*, 2003, TP.11.

Chapter 8
The MORPHEUS Data Communication and Storage Infrastructure

Fabio Campi, Antonio Deledda, Davide Rossi, Marcello Coppola, Lorenzo Pieralisi, Riccardo Locatelli, Giuseppe Maruccia, Tommaso DeMarco, Florian Ries, Matthias Kühnle, Michael Hübner, and Jürgen Becker

Abstract The previous chapter described the most significant blocks that compose the MORPHEUS architecture, and the added value they provide to the overall computation efficiency and/or usability. The present chapter describes the way that the memory hierarchy and the communication means in MORPHEUS are organized in order to provide to the computational engines the necessary data throughput while retaining ease of programmability. Critical issues are related to the definition of a computation model capable to hide heterogeneity and hardware details while providing a consistent interface to the end user. This model should be complemented by a data storage and movimentation infrastructure that must sustain the bandwidth requirements of the computation units while retaining a sufficient level of programmability to be adapted to all the different data flows defined over the architecture in its lifetime. These two aspects are strictly correlated and their combination represents the signal processor interface toward the end-user. For this reason, in the following, a significant focus will be given to the definition of a consistent computation pattern. This pattern should enable the user to confront MORPHEUS, in its strong heterogeneity, as a single computational core. All design options in the definition of the Memory hierarchy and the interconnect strategy will be then derived as a consequence of the theoretical analysis that underlines the computational model itself.

Keywords System-on-Chip • throughput • bandwidth • Petri-Net • Khan Process Network • Network-on-Chip • reconfigurable computing • DMA • dual port memory • DDRAM • memory controller

---

F. Campi (✉)
STMicroelectronics, Agrate Brianza, Italy
fabio.campi@st.com

M. Coppola, L. Pieralisi, R. Locatelli and G. Maruccia
STMicroelectronics, Grenoble, France

A. Deledda, D. Rossi, T. DeMarco, and F. Ries
ARCES – University of Bologna, Italy

M. Kühnle, M. Hübner, and J. Becker
ITIV University of Karlsruhe (TH), Germany

N. Voros et al. (eds.), *Dynamic System Reconfiguration in Heterogeneous Platforms*,
Lecture Notes in Electrical Engineering 40,

8.1 Computation Model

As introduced in Chapter 3, from the hardware point of view MORPHEUS is a heterogeneous multi-core System-on-Chip, where each computation unit (HRE) works as an independent processor connected in a Network-on-Chip. The final user should be in a position to partition its computation load utilizing for each application kernel the most suitable engine, while retaining a software-oriented approach to chip level synchronization and communication. The proposed programming model is organized at two levels: the first level is oriented at the global exploration of the application, its partition in specific concurrent kernels and the synchronization of the data flow and relative dependencies; this description will focus mostly at this level. A second level is oriented at the investigation of the most suitable computation fabric for each given kernel and the implementation of the kernel on the specific fabric making use of the fabric proprietary tools.

8.1.1 System-Level Organization: Micro-Code and Macro-Code

In principle, for the larger part of its applications, MORPHEUS is required to process data-streams under given real time constraints. Indeed, in their roughest and preliminary form, all application constraints and requirements are provided as bandwidth specifications.

The user is required to partition computational demands of the application over available hardware units and describe the relative communication patterns. In fact, the aim of the mapping task should be that of building a balanced pipelined flow in order to induce as few stalls as possible in computation, thus sustaining the required run-time specs. Hence, computation should be partitioned as much as possible on the three different HREs, and eventually the ARM core, in a balanced way (Fig. 8.1). Overall performance will be driven by the slowest stage, where a stage can be either computation or data transfer. Obviously, the timing budget of each stage is flexible, and can be refined by the user, much depending on the features of his application. The access point and interface between the user and all hardware facilities is the main processor core. Hardware resources are triggered and synchronized by software routines running on the ARM, either by manual programming or Real Time Operating System (RTOS).

The programming model is based on to the Molen paradigm [1]. The whole architecture is considered as a single virtual processor, where reconfigurable accelerators are functional units providing a *virtually infinite instruction set*. Tasks (i.e. application kernels) running on HREs or on ARM itself should be seen as instructions of the virtual processor. In order to manage the specificity of the HREs while preserving a homogeneous interface, the mapping of accelerations is library oriented: the user will have to either acquire a given library from the HRE vendor or develop it himself using HRE proprietary tools (see Chapters 4–6 and references [2–4]).

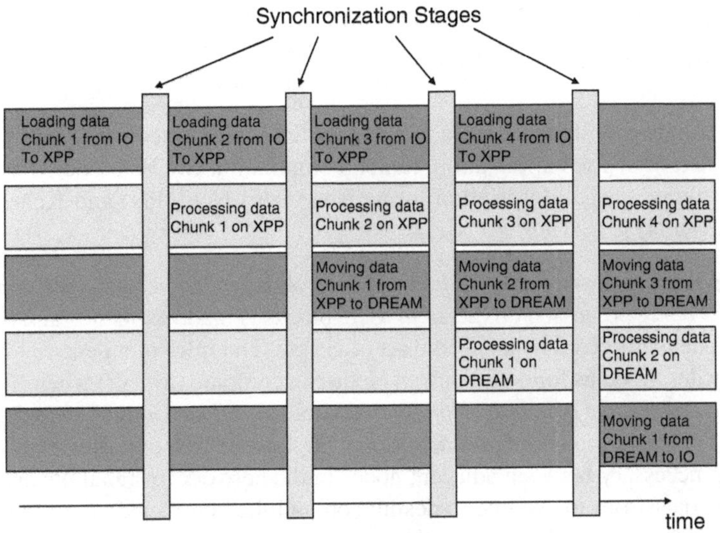

Fig. 8.1 MORPHEUS computational model

Bit-streams represent the virtual instructions micro-code, with the added value of being statically or dynamically reprogrammable. The work of compiler/RTOS and the configuration management is to schedule tasks in order to optimize computation, hide reconfiguration latencies, and ensure a familiar programming model to the user.

According to this paradigm, increasing the granularity of operators from ALU-like instructions to task running on HREs, we are forced to increase accordingly the granularity of the operands. Operands can not be any more scalar C-type data but become structured data chunks, referenced through their addressing pattern, be it simple (a share of the addressing space) or complex (vectorized and/or circular addressing based on multi-dimensional step/stride/mask parameters). Operands can also be of unknown or virtually infinite length, thus introducing the concept of stream-based computation.

From the architectural point of view we can then describe MORPHEUS handling of operands (source, destination and temporary data) at two levels:

- *Macro-Operand*, is the granularity handled by extension instructions, transferred by ARM and controlled by the end user through its main program written in C (possibly with the assistance of an RTOS). Macro-operands can be data streams, image frames, network packets or different types of data chunks whose nature and size depends largely on the application.
- *Micro-Operands* are the native types used in the description of the extension instruction, and tend to comply to the native data-types of the specific HRE entry language that is C for ARM, and C/GriffyC for DREAM, HDL for M2000, C/C++, FNC-PAE-Assembly and C/NML for XPP. Micro-operands will only be handled when programming the extensions, or macro-operators, so they are meant to be handled by the user only when for optimization reason he will program or modify manually extension operations on HREs.

8.1.2 Computation Formalism: Petri-Net, Khan Process Network

In order to preserve dependencies in the data flow without constraining too much size and nature of each application kernel, computation can be modeled according to two different design description formalisms: Petri Nets (PN) and Khan Process Network (KPN) [5].

- *Petri Nets*: Synchronization of data and control dependencies between application kernels being processed on different logic blocks is made explicit. Each computation node is triggered by a specific set of events. The rules of a generic PN can be briefly described as follows: A given Node can compute (trigger) when (i) *All preceding nodes have concluded computation* and (ii) *All successive nodes have read results of the previous computation.* Some implementation of hardware handshake is then necessary between adjacent nodes in the network to signal the availability of new data from the source to destination and the completed "consumption" of previously transferred data from the destination to the source of each transfer.
- *Khan Process Networks*: In this second case synchronization is modeled implicitly, by means of FIFO buffers that decouple the different stages of computation/ data transfer. There is no specific handshake between adjacent nodes. Each dependency in the system is modeled by means of an "infinite" FIFO that will hold data produced by the source until the destination is available for their consumption. In fact, KPN are mostly suited to "hardwired" implementation, since the actual dimensioning of FIFOs is very critical to avoid stalls, but that dimensioning is entirely related to the application. In a reconfigurable engine as MORPHEUS, the application of the KPN processing pattern may require a tuning of the kernels grain to match the granularity of FIFOs.

The choice of the most suitable formalism to be applied in each application deployment is related both to the features of the targeted application and to the nature of the targeted computational node. As it is possible to model a KPN through a PN but not the contrary, from the hardware point of view the PN model has been maintained as a reference although full support for KPN-oriented computation is maintained. The MORPHEUS architecture is a FULLY PROGRAMMABLE device: for this reason, each HRE must be configured before starting computation, and the configuration of a given node must be considered a "triggering" event for computation in the context of a Petri net. For this reason, a pure KPN pattern can not be applied, unless the configuration is totally *static* (i.e. all HREs and transfer are programmed only once in the application lifetime). In case of dynamic reconfiguration (i.e. the number of nodes in the KPN/PN is higher than the available HREs) KPN can be implemented as sub-nets, or second level nets of a larger PN triggered by the configuration events. In this case application nodes must be time-multiplexed and *scheduled* over available HRE.

Generally speaking, XPP appears suited to a KPN-oriented flow, as its inputs are organized with a streaming protocol. Unlike XPP, DREAM is a computation

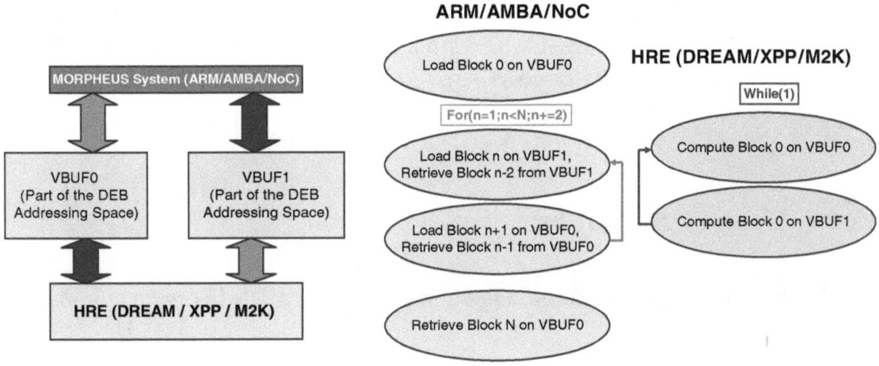

Fig. 8.2 Virtual Buffer "ping-pong" mechanism and FSM representation of the PN-based streaming computation concept

intensive engine: typically, input data are iteratively processed inside the DEBs, with the generation of a significant share of temporary results. Computation on DREAM thus can be more appropriately described as a collection of "Iteration Rounds" triggered by specific events, rather than a purely streaming computation. PNs appear more suitable to model the DREAM computation and its interaction with the rest of the system. Finally M2K is an eFPGA device, so that any computation running on it can be modelled according to either formalism depending on the RTL description of the eFPGA functionality. Local HRE memories (DEBs) and Cross-clock-domain exchange registers (XRs) represent the mean for providing the synchronization stage described in Fig. 8.1. The buffers can be utilized as FIFOs or as RAM memories. In the first case, a KPN pattern is realized, while in the second case it is necessary to guarantee data flow consistency with an explicit synchronization of data transfers and computation, in compliancy with the PN pattern (Fig. 8.2).

8.2 Implementation Details

As underlined in the introduction to this chapter, MORPHEUS aims at hiding the heterogeneity of the different storage and computation nodes in the architecture in order to provide a unified and homogeneous interface to the user.

This is obtained by describing the parallel MORPHEUS architecture as a sort of virtual sequential Von Neumann system, with ARM acting as control logic, and the HRE acting as function units. In this picture, the configuration bit-stream of each HRE acts as the instruction microcode.

As a consequence, the ARM core is utilized as one and only access point by the user to all utilities in the system. The status of the system is monitored by interrupts and consequent polling of the resources status registers (XR). All resources in the system (hardware services, computation nodes, storage nodes) are I/O mapped and any operation issued on the system is implemented by load/store operations by ARM, via bus and/or NoC access. Very often, when the control, data or configuration information to be transferred is large or complex the load/store operation is implemented as a DMA transfer.

8.2.1 Memory Hierarchy

As described in Chapter 3, communication in MORPHEUS is organized on three orthogonal layers:

1. CONTROL, SYNCHRONIZATION and DEBUG information is handled by the ARM processor by mean of the so-called "main AMBA [6] bus" (Fig. 8.3). All resources in the system are controlled by a set of registers located on this bus. Also, all on-chip and off-chip storage resources in the system can be accessed via ARM or DMA through this bus for debug purposes. The bus is hence critical but it is expected to carry only control information at computation time, so that bandwidth is not considered a significant issue.
2. CONFIGURATION BITSTREAMS for the various HRE are transferred on the so-called "Configuration AMBA bus". The bus can be controlled by ARM, DMA or by the configuration manager and also features off-chip access towards Flash/SRAM/DDRAM off-chip resources.

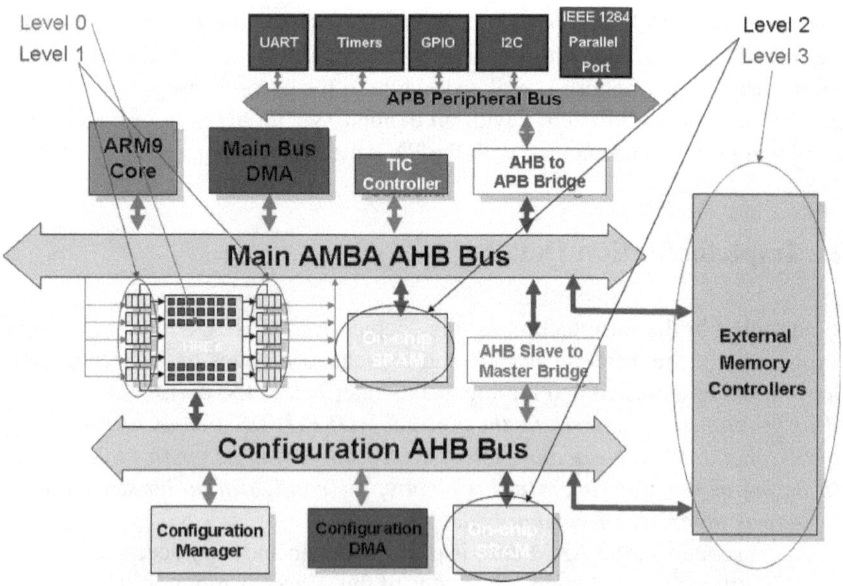

Fig. 8.3 MORPHEUS Bus Architecture and Memory Hierarchy caching layers. In the MORPHEUS architecture data transfer in implemented through the Network-on-Chip while the bus system is used to implement control and synchronization, HRE configuration, debug and testability access

3. APPLICATION DATA is transferred by a high-throughput NoC-based interconnect structure that allows direct access to external Flash/SRAM/DDRAM. Although it is possible for debugging purposes to transfer computation data by means of ARM or DMA transfers through the main bus, this option will not be applied during normal computation.

In order to implement the "balancing" between computation kernels and relative data transfers to fit in the "extended Heterogeneous pipeline" concept that is typical of MORPHEUS, each HRE is provided with local input/output buffers (DEBs). The same concept is applied to configuration exchange and system control, by means of configuration exchange buffers (CEB) and exchange registers (XR). All these buffers are implemented as dual clock, dual port memories. It may be argued that this option may impose area/power overheads, in the range of 40–80% depending on the target reference technology node. On the other hand, it ensures safe clock-domain crossing, and avoids complicated and costly multiplexing and synchronization logic on the memory ports.

These buffers serve also as a powerful mean to abstract the HRE heterogeneity and ensure consistent programming model and full scalability to the system:

- From the physical design point of view, the use of dual-port dual clock memories as only interface between the system and the computational kernel allows to design the specific accelerator without any information on the system it will be plugged in. Each HRE features its own clock control and generation domain and will be implemented on his own, as a predefined macro, easily reusable in different design contexts.
- From the architecture design point of view, it is possible to describe each accelerator as a triple entry in the addressing space of the system (Control/Configuration/Data) so that the only modification required by the introduction or removal of a given acceleration is the addition of a slave port to the configuration and control bus, and of a network node in the NoC-based data transfer infrastructure, so that scalability is strongly enabled both at the level of RTL implementation and/or System-C based design space exploration.
- From the programming model point of view, all HREs are seen as part of the addressing space, so that the user may work in the same way on any different configuration of the MORPHEUS platform. Assuming that most application kernels to be run on MORPHEUS should be either available as pre-packaged libraries (in case of standard functionalities) or produced by the MORPHEUS tool-set for very application-specific computations, the task of the final user would only be that of designing the data flow between the HREs as a balanced pipeline and ensure the data-flow consistency by appropriately programming the precedence between NoC transfers.

ARM features 16Kbytes I-cache and D-cache plus two 16Kbytes local software managed scratchpad memories respectively for Data and Instructions, referred to as Tightly Coupled Memories (TCM). All addressing not referred to TCM or cache is routed via AMBA master ports onto the system. ARM Cache and TCMs, as well as the DEBs and CEBs represent collectively the MORPHEUS L1 memory layer.

According to the information above, XPP DEBs are modeled as FIFOs, DREAM DEBs as processor controlled random access memories, whereas the DEBs connected to M2K are run-time programmable and support both access patterns. In order to exploit at maximum the hardware features of each HRE, the MORPHEUS is organized according to the Globally Asynchronous Locally Synchronous (GALS) clock distribution model and each HRE is provided with a local PLL that is programmable at run time by ARM. The size of DEBs ranges from 8 to 64Kbytes depending on the platform configuration. On-chip SRAM memory blocks are available as on-chip storage resources, partitioned between the main and configuration data bus. They are accessed by NoC or Bus access and represent the L2 caching layer, fully handled by software. Their sizes range between 100 and 500Kbytes depending on the platform configuration. The L3 storage layer is composed by off-chip memory. MORPHEUS features two different external memory control devices:

- A standard, asynchronous, 6-port 32-bit Flash/SRAM/ROM memory controller is used to access non-volatile support for storing boot-up, RTOS, system drivers and frequently used configuration bit-streams. The controller also allow debug of the on-chip bus system by standard AMBA-TIC (Test-Interface Control) verification protocol.
- An advanced, synchronous, 3-port 64-bit SDRAM/DDRAM controller, referred to as CMC was especially designed for sustaining the high data-bandwidth requirements that are expected form the MORPHEUS system. The CMC was described in detail in Section 7.2.

Both memory controllers are accessible by AMBA bus access (main bus/configuration bus). Flash and SRAM are expected to store Code, and Configuration information as well as low-latency access data. On the other hand, as it is supposed that data during peak computation will reside in DDRAM to ease I/O bandwidth, the SRAM controller is not connected to the NoC, while the DDRAM controller presents two dedicated NoC connections.

8.2.2 Data Interconnect Infrastructure

From many points of view MORPHEUS can be considered a multi-processor system-on-chip. Hence, the natural physical layer for implementing interconnects is a Network-on-Chip. The most relevant differences between MORPHEUS and a standard MPSoC environment reside in the size and granularity of HREs, which is quite different from that of standard processor acting as computation nodes in MPSoC:

HREs, differently from standard processors, feature clock speeds that strongly depend on the HRE granularity and also on the chosen application and the strategy deployed in its mapping.

1. HREs provide relevant computation capability: a network of HREs will feature a much smaller number of nodes but relevant bandwidth requirements for each transfer with respect to a network of processors. Also, data flows to/from HREs

are typically regular, rather then organized in bursts. As a result congestion issues will be less frequent, but they should be avoided at all costs as they could dramatically reduce computation efficiency starving computation nodes.

2. HREs usually feature long reconfiguration times. Hence, they tend to iterate the same computation kernel repetitively over large chunks of data. The configuration of the interconnect infrastructure can thus be defined as quasi-static: transfers are normally coarser than in a MPSoC environment, and the routing can be often maintained identical for a significant set of consecutive transfers generated by a given initiator node. As a consequence, a circuit switched communication pattern appears suitable. On the other hand, such approach may limit the flexibility of the communication and is more prone to congestion issues.
3. HREs are often based on streaming data access patterns or on automated regular addressing mechanisms. They can hardly behave as traffic initiators, because they rarely provide the addressing flexibility of a standard processor. Even in cases where this may be possible, the end-user would be forced to design and synchronize the system data flow programming different heterogeneous machines with inherently different languages and programming styles (e.g. C, HDL).

According to the points outlined above, the chosen approach to the deployment of the NoC concept in MORPHEUS is a network composed by few nodes, but with relevant bandwidth capabilities for each node-to-node connection. To avoid congestion, the topology (Fig. 8.4) was designed to provide direct and almost dedicated connections on the most critical paths. This notwithstanding, a circuit

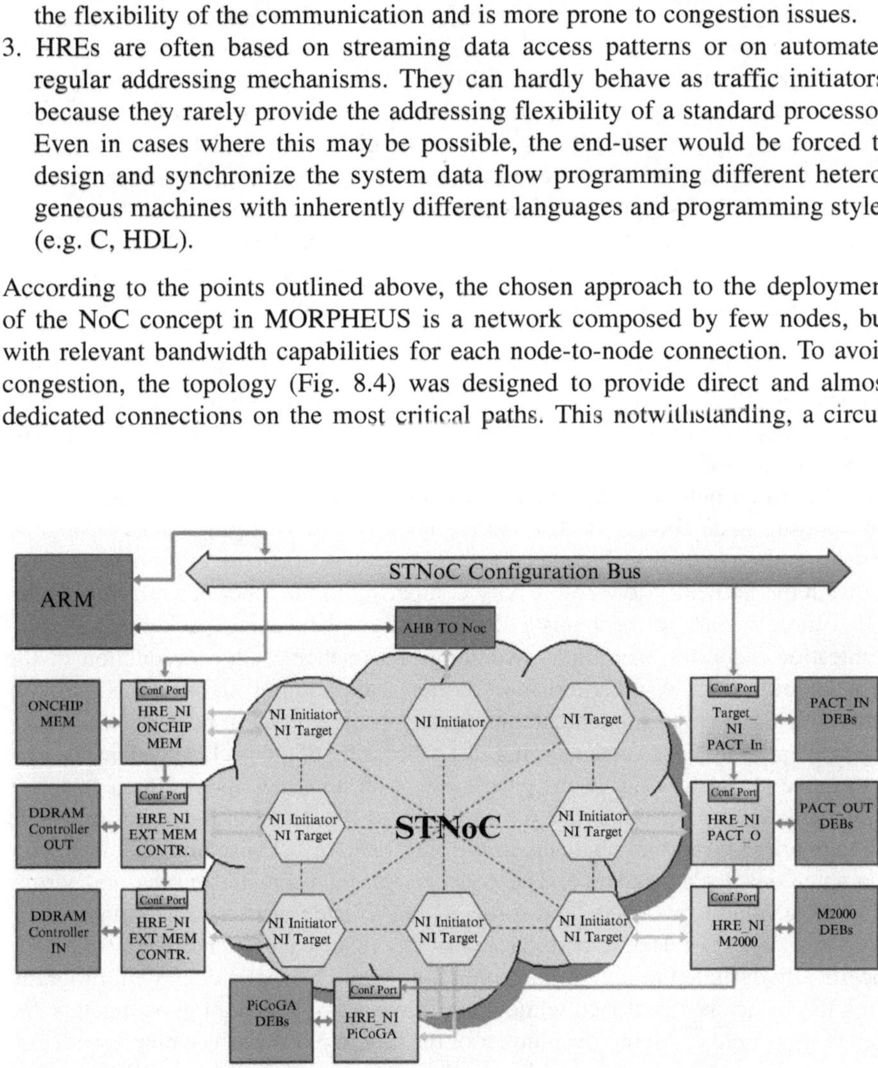

Fig. 8.4 MORPHEUS NoC topology and block description. See Fig. 3.2 for reference

switched approach appeared too application-specific, and did not provide sufficient guarantees for post-fabrication definition of new data-flows. A packet-switched net was preferred, where some paths were given absolute priority at design time but alternative minor connections are still possible and to a given extent can be eased altering at run-time priority schemes. Moreover, NoC-based MPSoCs tend to provide a distributed control, where many cores can act as traffic initiators. In the case of MORPHEUS, communication is strictly centralized and handled by the ARM processor.

To implement this feature without compromising parallelism MORPHEUS makes use of DMA engines distributed across the NoC initiator nodes, programmed by the ARM core, and performing transfer between HRE buffers (DEBs) and the architecture storage resources (SRAM, Memory controllers). Communication synchronization and control is handled by software routines running on the main processor, but this may impact performance and impose an awkward programming interface. As suggested in [7] this task can be performed by RTOS services, and part of the OS can be implemented in hardware: hardwired centralized control can be provided to support multi block transfers as well as of a synchronization concept and therefore releasing the processor/user from low-level tasks. From the application point of view, the programmer may then choose the preferred approach depending on constrains such as e.g. transfer data size or the activity level of the control.

The STNoC/Spidergon [8] concept was adopted as physical layer for the communication. Spidergon NoC is a environment for the design and implementation of Networks-on-chip based on a scalable, regular, point-to-point topology. The Spidergon network connects an even number of nodes in such a way that every node has three bidirectional connections: two to the neighbor nodes and one to the opposite node (Fig. 8.4). The main advantages of such topology are: regular structure, short network diameter (e.g. minimum amount of hops between two nodes in the network); low complexity comparing to the other network topologies and a simple routing mechanism. Spidergon implements a packet-based communication, adopting wormhole switching. This scheme allows reduction of the network buffering. A deterministic, shortest-path routing algorithm is utilized. The routing algorithm implementation is very simple, without posing the need for expensive routing tables and ensuring fast processing of the packet header. The idea is to move along the ring, in the proper direction, to reach nodes which are close to the source node, or otherwise to use the cross link to be in the opposite part of the network. Two virtual channels in the ring links (clockwise and anticlockwise) guarantee deadlock avoidance. The concepts of deterministic routing and virtual channel scheduling avoid costly disordered end-to-end transfers: the routing path of any packet does not depend on the route of any other packet. The NoC structure is specifically designed to hide implementation details, so that a consistent programming model can be developed without considering implementation parameters and may remain valid changing the number or the nature of HREs. The number of computation nodes and in particular the number of routers can be changed depending on architectural choices and floor-plan/timing analysis.

8.2.2.1 The Distributed DMA Concept

A NoC is by definition a distributed communication platform with a set of initiator nodes (e.g. processor cores) issuing transfers and a set of target storage nodes (e.g. memory units) responding to transfer requests. HREs represent peculiar nodes: they should be both NoC initiators (require transfers from some storage units such as on-chip or off-chip RAM), and targets (process external requests such as a transfer request from another HRE or ARM). This is implemented over the NoC through a distributed DMA pattern: in order to act as traffic initiator each HRE node Network.

Interface (HRE-NI, Fig. 8.5) is enhanced with a local DMA engine. HRE-NIs are thus capable to load data chunks from HREs and store them through the NoC to the target repository and vice-versa. From the user point of view this approach describes the whole NoC as an enlarged and highly parallel DMA architecture: ARM can require any transfer between HREs, as well as from any HRE to any storage unit (On-chip memory, Memory controllers). Transfers are initiated programming specific configuration registers on the HRE network interface, through a specific configuration bus reaching all HRE-NIs. NoC router priority schemes may also be programmed through the same configuration bus. As mentioned above, quasi static data transfers are expected to be the greatest volume of communication load on the network. This positively affects system control, performed by ARM by means of RTOS, since transfers can be issued as standard DMA transfers. On the other hand, since MORPHEUS is a single master system, it may suffer synchronization overload duc to an intensive interrupt activity (depending on the granularity of the tasks mapped on HREs). A possible solution is to map part of the RTOS services on hardware. A hardwired DNA (Data Network access) controller has been designed, featuring setup mechanisms for data transfers and a hardware handshake

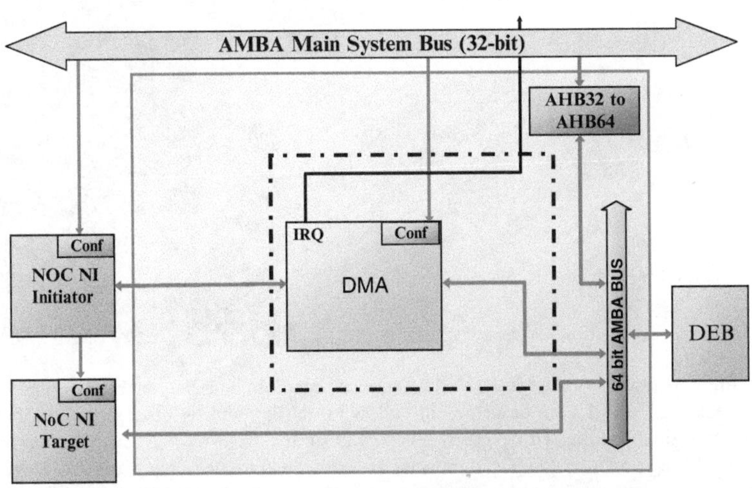

Fig. 8.5 Distributed DMA architecture: the HRE-NI concept

protocol that supports computation and communication synchronization and preserves data flow consistency. Its programming model is based on the same DMA interface used on the AHB control and configuration bus, so that it remains transparent for the end user. Eight concurrent programmable channels are supported, each capable to control multi-block transfers. Configuration items for channel setup can be programmed beforehand and related events will be detected and resolved by the DNA independently from further ARM commands. As a consequence, the user can design a given data flow according to three different approaches:

1. ARM act as "full-time" traffic controller. Code running on ARM monitors the status of each HRE through exchange registers (XRs) and triggers the required transfers over the HRE-NIs in order to maintain the desired stream through the system. This is very useful in the first stages of application deployment, to evaluate the cost of each step in the computation, maintain full programmability, and check for bottlenecks.
2. ARM acts as "batch" controller and enabler. After a "configuration" phase in which ARM configures all HREs and relative transfers on the HRE-NIs, it remains waiting for interrupts. This approach is necessary in case of a *controlled* computation network (application that requires dynamic reconfiguration to schedule different PN nodes over the same HRE) or in any case the user may prefer to deploy a PN.
3. The network can be self-synchronized: ARM provides the initial configuration phase, and after that the HRE-NI will iterate over regular addressing implementing

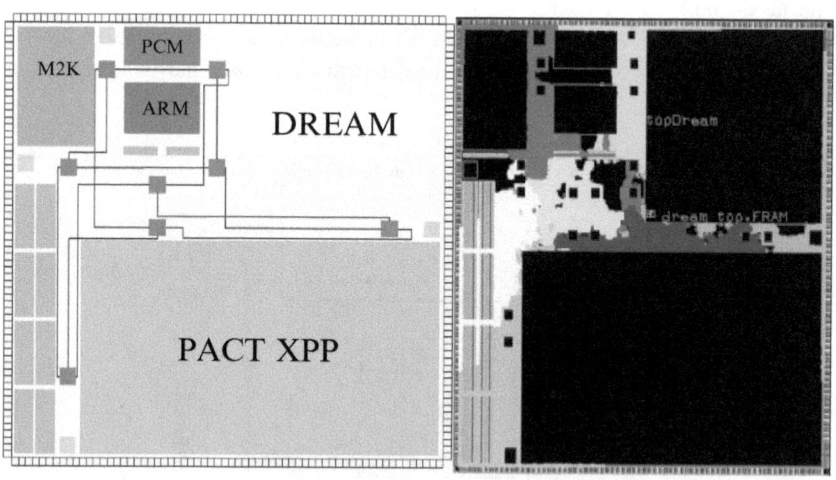

Fig. 8.6 Implementation of the MORPHEUS Network-on-Chip on a reference Test-chip: Planned (*left*) and real (*right*). PLLs are described in yellow on the *left* and by black boxes on the *right*, routers are described on the *left* by green boxes, and on the *right* by pairs of black boxes (one router per each ring direction). DPDC memories are embedded in the relative macros. The colored areas on the *right* represent network interfaces of M2K, dream, ARM, XPP In and Out, Onchip Memory, DDRAM Controller In and Out

a fixed data-flow through the system. This can be deployed for static applications or, most likely, for a limited time-share of the application as a second level KPN included in a larger PN network.

References

1. S. Vassiliadis, S. Wong, G.N. Gaydadjiev, K. Bertels, G.K. Kuzmanov, E. Moscu Panainte, The Molen Polymorphic Processor, pp. 1363–1375, *IEEE Transactions on Computers*, November 2004.
2. PACT XPP Technologies, PACT Software Design System XPP-IIb (PSDS XPP-IIb) – Programming Tutorial, Version 3.2, www.xpp.com, 2005.
3. M2000, Flexeos Software User Manual, Version 2.4.4, www.m2k.fr, 2006.
4. C. Mucci et al., A C-based Algorithm Development Flow for a Reconfigurable Processor Architecture, IEEE SOC, Tampere, 2003.
5. G. Gao, Y. Wong, Q. Ning, A Timed Petri-Net Model for Fine-Grain Loop Scheduling ACM SIGPLAN June 1991.
6. AMBA Specification (Rev 2.0), ARM Ltd 2001.
7. T.A. Bartic et al., Topology Adaptive NoC Design and Implementation, *IEE Computers and Digital Techniques*, July 2005.
8. M. Coppola et al., Spidergron: A Novel On-chip Communication Network, IEEE SOC, Tampere, 2004.

Part III
The Integrated Tool Chain

Chapter 9
Overall MORPHEUS Toolset Flow

Philippe Millet

Abstract The obvious complexity of a heterogeneous architecture like MORPHEUS is hidden to application designers thanks to an integrated set of tools. Deploying an application is as easy as writing C code, and assembling IPs.

Keywords Code generation • data movements and parallelism

9.1 The Need for a Toolset

One way, that became fast the well-known classical and usual way, of developing an industrial application is to write a linear C code for a single core processor. This way of working fits well the mono-core computing architectures to which we have had a huge experience during the last 30 years and introduced a huge legacy in the industry.

The MORPHEUS platform intends to have an innovative architecture approach with three heterogeneous reconfigurable computing accelerator units called Heterogeneous Reconfigurable Engines (HRE). This heterogeneity gives not only a very flexible solution that allows designers to implement application from several domains on the same chip, but also a solution that can adapt its processing to environment constraints at real-time.

The drawback is an increase of programming complexity. Implementing an application on the architecture require the use of the three accelerator languages as well as understanding data exchanges and synchronisation mechanisms that take place between computation units.

The embedded market targeted by MORPHEUS is very sensitive to productivity, code reusability (e.g. legacy application) and programming efficiency. Such a complex architecture would not get any interest from the embedded community if it would not be possible to easily program it.

P. Millet (✉)
Thales Research & Technology, France
philippe.millet@thalesgroup.com

N. Voros et al. (eds.), *Dynamic System Reconfiguration in Heterogeneous Platforms*, Lecture Notes in Electrical Engineering 40,

To be able to assist the user during implementation phases of his project, the MORPHEUS platform provides a toolset that takes care of the complexity of the architecture and abstracts the heterogeneity of the accelerators.

9.2 The Objectives of the Toolset

The ambition of the MORPHEUS platform is to combine the best from General Purpose Processors (GPP), Application Specific Integrated Circuits (ASIC), and Field Programmable Gate Arrays (FPGA). Getting simplicity of programming from GPPs, ASICs' low power consumption, and FPGAs' flexibility.

The objective is to have a simple design flow of applications as easy to write as a C code. Thanks to the MORPHEUS toolset, designing an application remains very close to the well-known classical way. The toolset hides all the complexity and helps in deploying applications on the architecture (Fig. 9.1).

When implementing an application on the architecture, the designer has to split its application into scheduling and control parts and into computing intensive parts. The control parts are given to the ARM CPU, while the computation functions are given to the HREs.

Many synchronisations and communications have to take place between the ARM and the HREs. Mainly because the ARM controls data transfers in the architecture, and because the memory in the HRE (DEB) is small in comparison to the usual application needs.

Three HREs are included in the MORPHEUS chip. They give the application designers opportunities to execute their accelerated functions in parallel. Two kinds of parallelism are considered: (1) dataflow instantiated as a static pipeline chain and (2) and shared memory (Fig. 9.2).

The static pipeline chain is built during the implementation of the accelerated function. The designer can split an accelerated function on up to the three HREs. The toolset synchronises the HREs and uses the data exchange descriptions to move the data from the output of one HRE directly to the input of the next HRE in the pipeline.

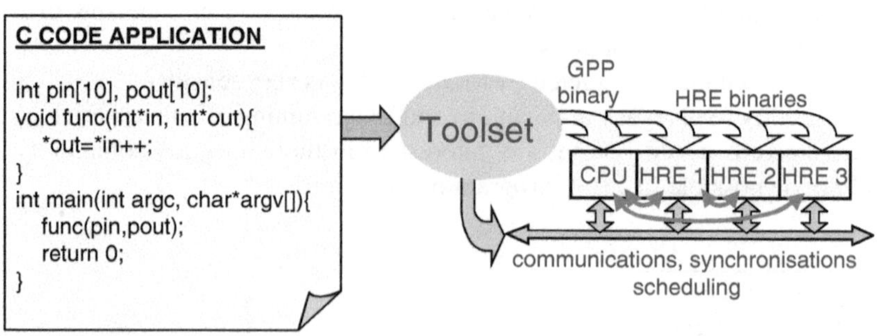

Fig. 9.1 Toolset objectives

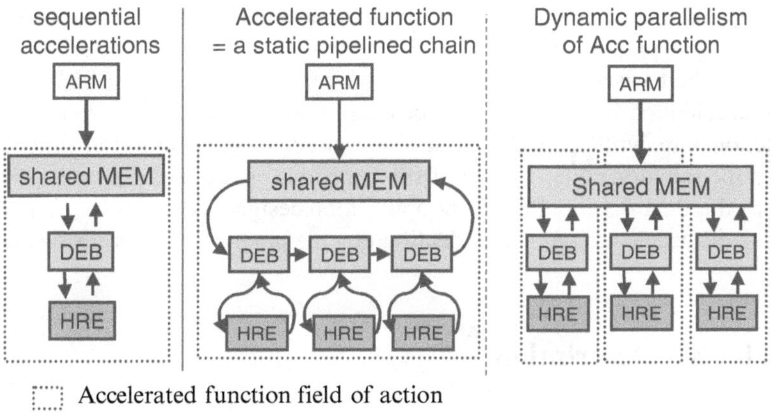

Fig. 9.2 Parallelism supported by toolset

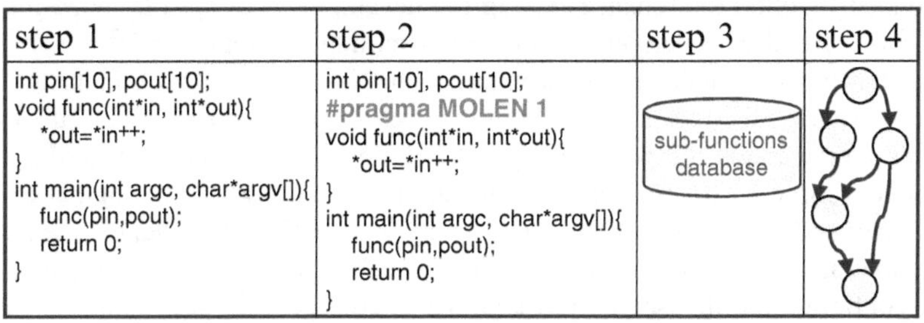

Fig. 9.3 Programming steps

With the shared memory scheme, the HREs are used in parallel on the same set of data with either different accelerated functions on each HRE or on three different implementation of the same function on the three of them.

Both cases are handled by the toolset and any combination is possible (e.g. two HREs for a pipelined function and the last HRE for another function running in parallel).

The flexibility of the architecture to reconfigure parts of its computational units gives the platform the ability to find the best scheduling at each execution time. When an accelerated function is called, the platform selects the best accelerator for executing this function, and schedules the total set of functions to get the best performances.

To hide the complexity of parallelism and heterogeneity, the toolset need the user to program its application with the steps from Fig. 9.3.

1. The application is written in standard C code. This gives the user the opportunity to use powerful standard development tool chains. During this step, one can validate the application against use cases and test benches.

2. The application designer identifies for the toolset the functions to be accelerated by setting a pragma on top of each of them in its initial C code. Using this pragma, the toolset knows which function runs on the HRE.
3. The accelerated functions are composed of elementary sub-functions. The sub-functions are written in C. The toolset extracts input and output information from the function code. The set of sub-functions creates a database.
4. Using this database as a LEGO, the application designer connects sub-functions to create the accelerated function identified by the pragma in the original C code.

9.3 Toolset Description

To achieve its missions, the toolset is composed of several tools which are grouped in three categories: (1) an operating system handling the scheduling of the whole application and delivering services to the application for accessing the architecture, (2) an accelerated function tool chain handling the development of specific code for each HRE, and (3) a ARM development tool chain handling the control and the overall data movements in the architecture for the application. The number of tools in the toolset is not an issue since they are hidden. From the user point of view (Fig. 9.4), the toolset remains simple and is composed of a graphical tool (SPEAR) and a makefile.

The user input for the toolset are (1) the application C code with pragmas to identify accelerated functions, (2) the database of sub-functions C code, (3) the schema giving interaction between sub-function to create the accelerated functions and (4) an XML configuration file used by the compiler to configure the operating system.

This simplified interface hides tools cooperating on the tasks of the toolset which are (1) analyse of the user sub-functions, (2) generate data movement and synchronisation when needed, (3) generate HRE binaries (or bitstreams), (4) schedule the whole application by loading the HRE in time for execution and launching the DMA for data movements, (5) compiles the sets of binaries and make a MORPHEUS package, and (6) deploy and run the application on the architecture or on a simulator of the architecture. The Fig. 9.5 shows the interconnection between tools.

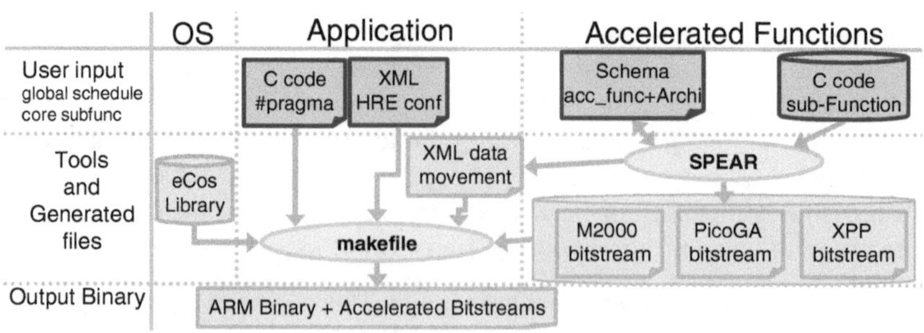

Fig. 9.4 Toolset user's view

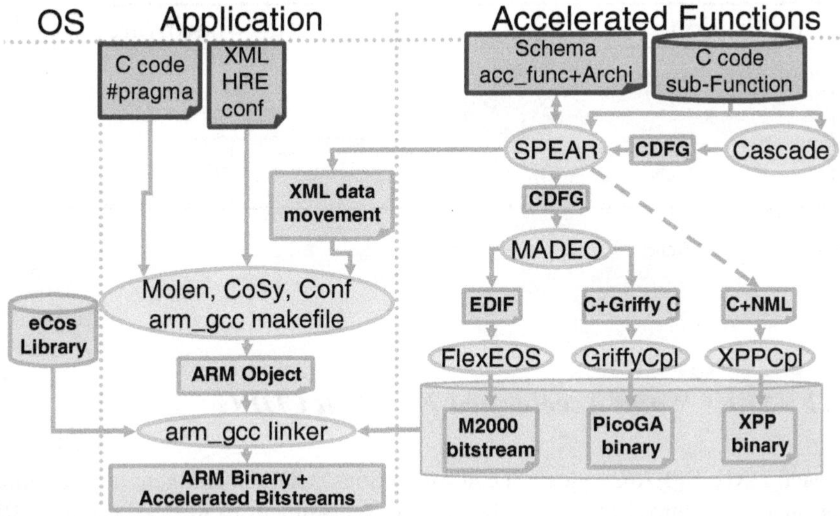

Fig. 9.5 Links between tools within the toolset

To understand what takes place within the toolset and how tools are cooperating together, one has to understand each tool.

9.3.1 Design an Accelerated Function in SPEAR

For each accelerated function the user creates a project in SPEAR. This tool needs the user to describe both his *application* (software) by linking sub-functions from the database, and his *architecture* (hardware) that runs the application. Each function is described as a chain of sub-functions. Each sub-function is mapped on a HRE. Therefore a pipeline is described by mapping each sub-function on a different HRE. Several sub-functions can also be mapped on the same HRE. In the given example an accelerated function is composed of three sub-functions (Fig. 9.6).

9.3.2 CASCADE and Sub-function C Code Analysis

SPEAR calls CASCADE to analyse each sub-function and convert them from C to a Control Data Flow Graph (CDFG) using the Standard for the Exchange of Product model data (STEP[1]) format. CDFG is more comfortable for code analysis and modification than C Code. The file generated by CASCADE describes the core of the elementary sub-function. At this stage the file does not contain data movements information. The result from cascade is a set of CDFGs, one per sub-function.

[1] STEP is an ISO- 10303-21 standard.

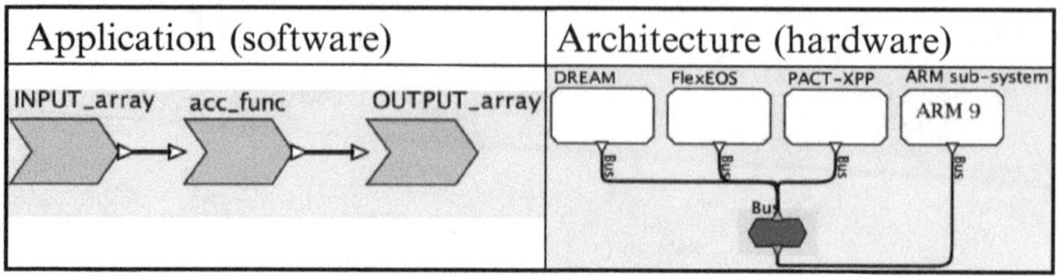

Fig. 9.6 SPEAR schema

9.3.3 SPEAR and Accelerated Function CDFG

For each HRE, SPEAR collects the sub-functions' CDFGs and generates communication and synchronisation code between the sub-functions mapped on this HRE. It then creates a single CDFG containing the whole part of the accelerated function mapped on the HRE.

9.3.4 SPEAR, HREs Communication and Shared Memories

When the sub-functions are mapped on the HREs, SPEAR uses the information about inputs and outputs of the sub-functions to generate configuration of data movement in the architecture, not only between shared memories and each HRE, but also between HREs.

For each input (each output) of the part of the accelerated function mapped on each HRE, SPEAR creates a description of how to feed that input (that output). Figure 9.7 gives a sample of memory movement description.

This description allows the tool to describe any movement with any number of dimensions. In this example we want to process a full picture that does not fit in HRE's memory. The picture is divided into sub-pictures, i.e. a two-dimensional set of data. A DMA block is a continuous set of data in memory i.e. a single dimensional set of data. A list of DMA blocks has to be given to the DMA in order to load the whole sub-picture in the HRE. To find each block of a sub-picture, the state machine adds the step of dimension 1 to the current memory offset. In this example, four DMA blocks are needed to load a sub-picture. After a sub-picture is loaded, the HRE starts processing. The next sub-picture to be loaded is at an offset given by the step of dimension 2 from the origin of the full picture. The next block is found at this current position incremented by the step of dimension 1, so again a complete dimension 1 cycle of four blocks is played. When the dimension 2 cycle is finished, the next block is found by adding the step of dimension 3. And again dimension 2 and dimension 1 cycles are played. Up to processing completely the full picture. The same description is done for the output.

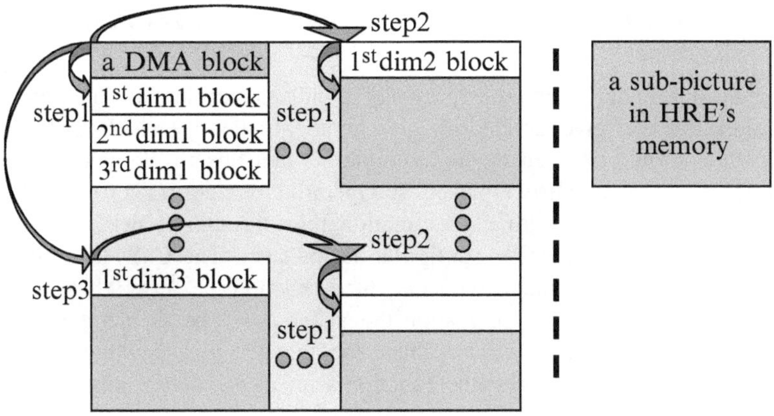

Fig. 9.7 Memory movement description file example

Few DMA can deal with two dimensions and very few can handle more, the method presented here, is suitable for any DMA, the driver takes into account the number of dimensions the DMA can manage. The diver software manages the remaining dimensions.

With more than one input (or output) on an accelerated function, a scheduling of the inputs/outputs has to take place in order to start the processing in the HRE when the input is ready and when the output is free. Such a scheduling is also necessary when one wants the data movement to overlap with processing in order to reduce the latency. Such a description is also generated by SPEAR and taken into account by the data movement state machine in the operating system.

At the end of its generation process, SPEAR calls MADEO and the HREs compilers to generate the binaries and delivers the XML description of the data movements to MOLEN tools.

9.3.5 MADEO and HRE's Compilers for HRE's Binaries

MADEO generates the HRE's source code from the CDFG generated by CASCADE and SPEAR. Thanks to the information stored in the CDFG, the generated code for the HREs exactly matches the data transfers taking place in the architecture. The synchronisation mechanisms are also automatically generated. The user does not have to know how to program the HREs, he does not have to take care of the data movements between the external memories and the HRE's memories, and does not have to deal with the synchronisations between the cores, all this is generated from the information stored in the schema.

The compiler of each HRE is called by SPEAR to generate a database of binary accelerated functions. This database together with the XML description file for data movements are given to the ARM tool chain. In this database, each accelerated function has its own identification number.

9.3.6 ARM Tool Chain, MOLEN, CoSy Compiler and eCos

The application main C code uses pragmas to indicate accelerated functions. With this pragma, the user gives an identification number to identify which function from the database should be used by the compiler. When this accelerated function is called, the C function call is replaced by Operating System (OS) calls (Fig. 9.8). The OS is based on eCos with added functionalities through MORPHEUS drivers, data movement manager, HRE configuration manager. The CoSy compiler calls these functionalities to handle the accelerated functions. In Fig. 9.8 a sample of application is given. In this application the_acc_func is the accelerated function. The MOLEN_FUNCTION pragma above the_the_acc_func declaration tells the compiler that this is an accelerated function with identification number 4 in the database.

This function is called in the test_picoga function. The generated assembly code from the CoSy compiler test_picoga shows that the C call to the_acc_func is replaced by a set of OS function calls to handle the DMA configuration, load the bitstream in the HRE, start processing with the HRE, wait for the job to be finished, and saying the HRE this accelerated function is no longer necessary. During DMA configuration, the OS create a thread that manages the scheduling of the data movements and the HRE processing.

The function in charge of DMA configuration and called in assembly code is generated by a MOLEN pre-processor that understands the XML file from SPEAR and configures the data movement state machine.

The ARM compiler embeds the binary (or bitstream) of each accelerated function in an array mapped in the configuration memory. This array is loaded into the HRE during the molen_SET function call. Several accelerated functions can be launched in parallel. A management thread with a dedicated state machine is created for each such accelerated function.

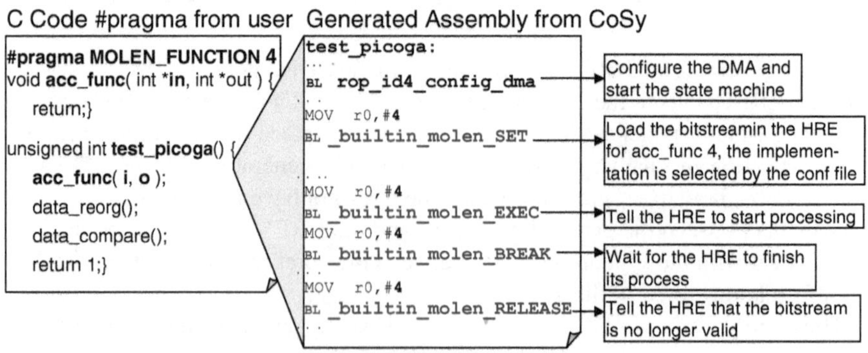

Fig. 9.8 Accelerated function call

9.4 Conclusions

In this chapter we have described the toolset developed for the MORPHEUS platform. From C code description the toolset generates accelerator target specific codes and optimises the schedule of the application.

Chapter 10
The Molen Organisation and Programming Paradigm

Koen Bertels, Marcel Beemster, Vlad-Mihai Sima, Elena Moscu Panainte, and Marius Schoorel

Abstract This chapter presents the Molen organization which is used as the way the processor and the reconfigurable devices will interact. Based on this organization, the Molen programmer's interfaces are presented and the way the compiler schedules the configuration and execution instructions is presented. We also present potential extensions based on OpenMP for parallel execution of the hardware kernels.

Keywords Computer organisation · compiler · parallel execution

10.1 Introduction

The idea of adding accelerators to a general purpose processor is certainly not new. Floating point units are a classic example of such an accelerator in the first generations of microprocessors. In today's platforms, we are increasingly confronted with heterogeneous cores and varying requirements that may change not only during design time but also at runtime. It is especially in such a context that reconfigurable cores are interesting. However, one of the challenges is to define a transparent way to combine the general purpose processor with the reconfigurable accelerators. The Molen organization and the corresponding programming paradigm provide a solution to most of the issues that arise such as:

- *Opcode explosion*: one could envision that each application specific instruction would be mapped on a unique opcode. However, given the large applicability of such reconfigurable cores, this would rapidly exhaust the available opcode space.

K. Bertels (✉) V.-M. Sima, and E.M. Panainte
Computer Engineering Lab, Delft University of Technology, The Netherlands
k.l.m.bertels@tudelft.nl

M. Beemster and M. Schoorel
ACE BV, The Netherlands

N. Voros et al. (eds.), *Dynamic System Reconfiguration in Heterogeneous Platforms*,
Lecture Notes in Electrical Engineering 40,

- *Limitation of number of parameters*: in some cases, the number of parameters is limited because of encoding limitations. However, this imposes a very rigid restriction on the reconfigurable core.
- *Lack of parallel execution*: one of the most powerful features of reconfigurable cores is the capability to execute in parallel many operations. If the architecture does not support this, it severely limits the impact of using such cores.
- *No modularity*: as platforms evolve, we need to be able to port existing applications easily to the latest one. A modular approach will ease such porting as additional will be mostly restricted to the interfaces and the actual processing logic inside can be maintained.

10.2 Molen Organisation

The Molen organisation provides an answer to the above challenges by providing a one time extension of the instruction set combined with a microcoded emulation of complex operations as a sequence of simpler and smaller basic operations. The one time extension of the instruction set, controlled by the reconfigurable microcode, consists of the following:

- *SET <op>*: starts the configuration of the reconfigurable core.
- *EXECUTE <op>*: starts the actual execution of the reconfigurable core.

These two basic operations guarantee that arbitrary functionality can be implemented. The organization that supports this is depicted in Fig. 10.1. Instructions

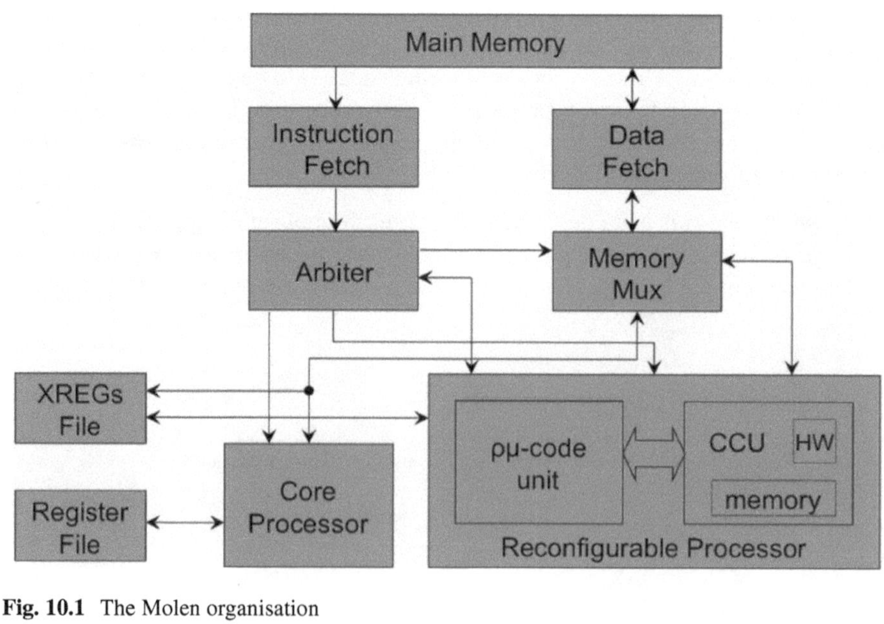

Fig. 10.1 The Molen organisation

are issued to the arbiter who determines whether the instruction is for the reconfigurable core or the general purpose processor. Data is fetched by the Data Fetch unit and through the Memory Multiplexor dispatched to either the reconfigurable core or the general purpose processor. The reconfigurable processor consists of the reconfigurable micro code unit and the custom configured unit (CCU). We refer to [1] for a detailed description of the Molen polymorphic processor.

The Set and Execute instructions provide the programmer with the necessary interfaces to be able to indicate in the source code what parts of the application are mapped on the accelerator and what part will be executed by the general purpose processor.

However, a number of issues still need to be resolved before the anticipated speedup can be achieved. As the time necessary to configure the CCU is in the order of milliseconds, it is of vital importance to be able to hide this latency as much as possible. This is done by scheduling the Set instruction at a point in the execution flow of the program such that the configuration is completed at the moment when the Execute statement is received from the instruction memory. However, there is also a spatial aspect: each CCU will occupy area on the reconfigurable device and we may need to place multiple CCU's at the same time. Again, we have to identify which CCU's need to be mapped at one point in time and whether enough area is available.

In the context of the MORPHEUS project, the Molen organization consists of an ARM processor to which are connected various reconfigurable devices, namely the PicoGA, the M2000 and the Pact XPP. Whether they are interconnected by means of a NoC, buses or dedicated communication channels are more implementation issues than conceptual differences.

10.3 Molen Compiler

As with any particular computing platform, we need to have a compiler that understands the hardware platform and is capable of exploiting it. In this respect, we present the compiler back end related issues that have been implemented using the CoSy compiler platform.

In our discussion, we will refer to the 'simple' Molen architecture where the gpp is linked to one FPGA. However, this can be extended to any specific hardware platform as long as the shared memory, tightly coupled processor co-processor template is adopted.

In order to guarantee that the required speedup[1] will be achieved, the compiler will have to make a temporal and spatial schedule of all the necessary Set/Execute statements. In case no feasible schedule can be found, the compiler will revert to the original source code.

Before presenting the dataflow equations that will take care of this, we briefly discuss the pragma annotations that need to be inserted to indicate to the compiler what parts will be mapped on the reconfigurable device.

[1] We assume speedup to be the design objective. It could also be low power, low bandwidth, etc.

10.3.1 Pragma Annotations

In Table 10.1, we find the original C-code where a main function calls the function **add**. The function declaration is preceded by a pragma annotation which informs the compiler that a call to an FPGA is requested. As indicated above, the compiler will then insert at the intermediate code level the necessary Set and Execute instructions (together with some additional instructions to move the parameters to dedicated exchange registers). In this simple example, the Set basically precedes the Execute statement which in reality would be scheduled earlier such that the latency of the configuration is hidden.

In principle, this pragma annotation is the only thing the programmer has to do, assuming of course that the hardware kernel (**add** in this case) is available.

10.3.2 Spatio-temporal Scheduler

The spatio-temporal scheduler has as inputs: the control flow graph with MOLEN instructions and information regarding potential conflicting hardware requirements between various operations (a conflict occurs when two hardware kernels can't be configured at the same time, because they use the same resources). Through profiling, we annotate the edges of the control flow graph with additional information, namely, the execution frequency of that edge. This information can be used to make a choice between conflicting hardware requests such that the compiler picks the CCU which is executed most. A detailed description of this scheduler is given in [2].

10.3.2.1 The Anticipation Subgraph

Partial anticipability – a hardware operation **op** is partially anticipated in a point m if there is at least one path from m to the exit node that contains a definition node

Table 10.1 Pragma annotations and intermediate code

#pragma call_fpga add	Main:	mrk 2, 14
int f(int a, at b) {	mrk 2, 13	mov $vr2.s32 <- main.z
int c;	ldc $vr0.s32 <- 5	movtx $vr1.s32(XR)<-$vr2.s32
c=a+6;	mov main.z<-$vr0.s32	ldc $vr4.s32 <- 21
return c;	mrk 2, 14	movtx $vr3.s32(XR)<-$vr4.s32
}	ldc $vr2.s32 <- 21	**set** (add)
	call $vr1.s32<-	ldc $vr6.s32<-0
void main() {	f(main.z,$vr2.s32)	movtx $vr7.s32(XR)<-$vr6.s32
int x,z;	mov main.x<-$vr1.s32	**exec** (add) $vr5.s32(XA) <-
z=5;	mrk 2,15	$vr1.s32(XR), $vr3.s32(XR)
x=f(z,21);	ldc $vr3.s32<-0	movfx $vr8.s32 <- $vr5.s32(XR)
}	rer $vr3.s32	mov main.x <- $vr8.s32
	.text_end main	
C code	**Intermediate code**	**Modified intermediate code**

for op and none of the paths from m to the first such definition contains a conflict node for op.

The nodes in which an operation **op** is partially anticipated are the nodes in which that operation can be configured and still be available at the execution point (as no conflicting operations are from that node to the execution node).

$$\text{PANTin(i)} = \text{Gen(i)} \cup (\text{PANTout(i)} - \text{Kill(i)})$$
$$\text{PANTout(i)} = \bigcup_{j \in \text{Succ(i)}} \text{PANTin(j)}$$
$$\text{PANTout(exit)} = 0$$

The generation of this graph is a backward dataflow problem. The equations are the standard equations used for such an analysis with the exception of the union operation used to compute PANTout(i). The first equation 'propagates' back what is available at the output, excluding the operations that are in conflict with the operations generated in this node (this is the set Kill(i)) and adding the operation generated at this node.

The union in the second equation is a special union that excludes conflicting hardware operations, and it is defined as:

$$A \bigcup B = \{x \in A \bigcup B \,|\, \neg \exists y \in A, x \xleftrightarrow[\text{conflict}]{} y\}$$

10.3.2.2 The Availability Subgraph

This graph is used to eliminate the hardware configuration when they are already available. This is a forward data flow problem with the following equations:

$$\text{AVALout(i)} = \text{Gen(i)} \bigcup (\text{AVALin(i)} - \text{Kill(i)})$$
$$\text{AVALin(i)} = \bigcap_{j \in \text{Pred(i)}} \text{AVALout(j)}$$
$$\text{AVALin(entry)} = 0$$

The **Kill** and **Gen** sets are defined the same as for the previous graph – the Gen are the operations configured in this node, and the Kill are the operations in conflict with the ones generated.

The results of the two dataflow analyses are presented in Fig. 10.2.

10.3.2.3 The Anticipation Graph

Based on the previous subgraphs, for each operation **op**, we eliminate from the initial graph the nodes which are not essential. An edge (u,v) is considered

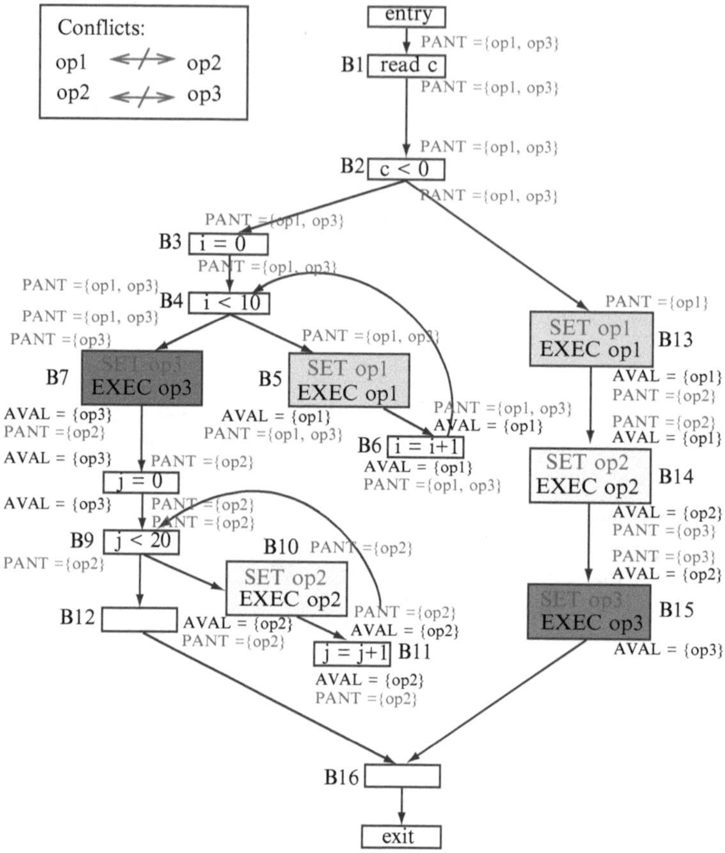

Fig. 10.2 Partial anticipability and availability analysis

essential (and thus, not eliminated) if $op \notin AVALout(u) \wedge op \in PANTin(v)$. The reduced graph contains all the nodes that are in at least one essential edge and all the essential edges. New pseudo entry and exit nodes are introduced and edges connect any node that doesn't have a predecessor to the entry and any node that doesn't have a successor to the exit. This new edges have an infinite execution frequency (Fig. 10.3).

10.3.2.4 Minimum Cut

For each operation, the minimum cut is applied on the anticipation graph. The edges contained in the cut are the edges on which the set operations can be introduced while minimizing the number of reconfiguration. As each reconfiguration is time consuming, minimizing the number of reconfiguration will minimize the total execution time.

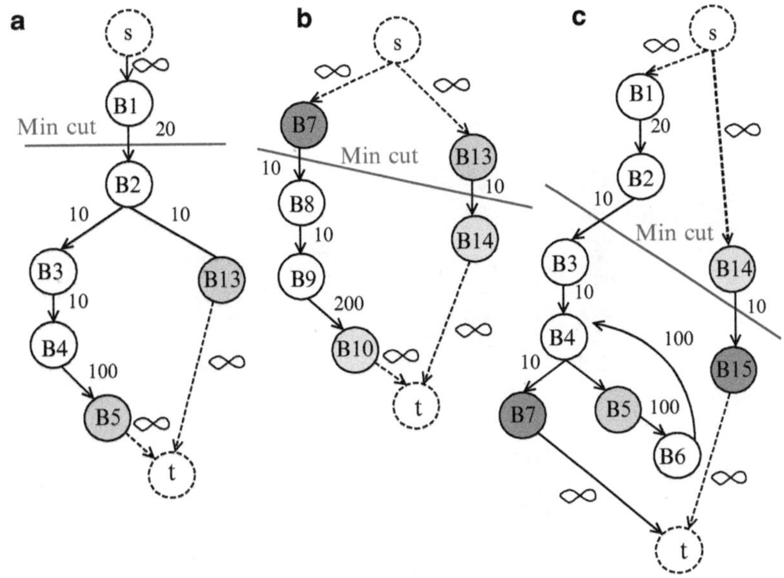

Fig. 10.3 Anticpation graphs for (a) op1, (b) op2 and (c) op3

10.4 OpenMP Extensions for Molen

Running Molen in a multi-application/multi-threaded environment introduces new challenges. One of these challenges is allowing the operating system to control the execution and scale the execution of hardware kernels.

OpenMP is set of compiler directives, library functions and environment variables that can be used to specify parallelism in applications developed for shared memory architecture. These characteristics make OpenMP an obvious choice for specifying parallelism in an application that uses Molen programming paradigm (Fig. 10.4).

An example of OpenMP code is the following:

```
int func(int **in, int **out, int l)
{
int i;
#pragma omp parallel for
for(i=0;i<l;i++)
{
#pragma molen 1
  filter(in[i],out[i]);
}
}
```

The code contains two annotations: the first one, "*#pragma omp parallel for*" is a pragma defined by the OpenMP standard to mark the iterations of the next loop can be executed in parallel (there are no data dependencies between them).

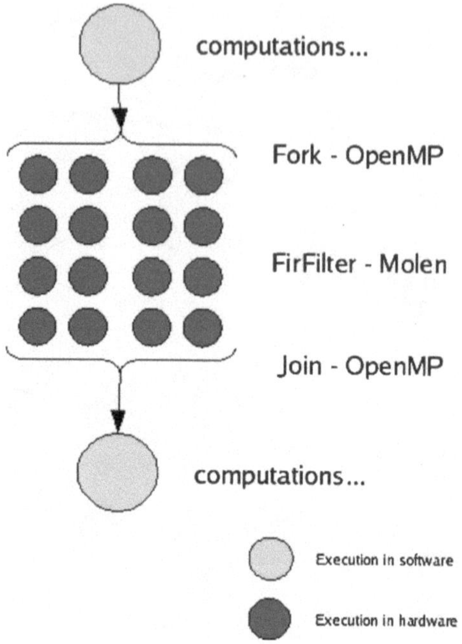

Fig. 10.4 Molen in the context of OpenMP

The MOLEN pragma was described in previous chapters.

The OpenMP philosophy is that you shouldn't specify the number of threads, but to let the runtime system to decide how many threads should be created.

Assuming the implementation fits only once on the available reconfigurable fabric, we need to use just one thread so the code generated will be equivalent with the code generated for the C program without the OpenMP pragma.

Obviously in case we have several threads we can't use the above code as for each thread another instance of the implementation has to be run. So, instead of using 'instructions' we will use MOLEN operating system calls. For each thread we will create a handle that will represent a set of hardware instances – the ones the current thread has to use. This case is depicted in Fig. 10.4.

The generated code will look like:

```
int func(int **in, int **out, int l)
{
int ops[]={1};
long i, _s0, _e0;
if (GOMP_loop_runtime_start (0, n, 1, &_s0, &_e0))
  do {
    h = molen_create_operations(ops);
    for (i = _s0, i < _e0; i++) {
```

```
      molen_movtx(h,1,0,in[i]);
      molen_movtx(h,1,1,out[i]);
      molen_execute(h,1);
      molen_break(h,1);
    }
  } while (GOMP_loop_runtime_next (&_s0, &e0));
  molen_destroy_operations(h);
GOMP_loop_end ();
}
molen_destroy_operations(h);
GOMP_loop_end ();
}
```

For each pair (handle, id) there will be a hardware instance available, and the molen_ functions will call each one from the threads.

10.5 Conclusions

In this chapter, we have introduced the Molen organization which has been adopted in the MORPHEUS project and discussed how the general purpose processor and the reconfigurable accelerators can interact. We also presented how the developer can make use of the heterogeneous platform by introducing simple pragma annotations and we explained how the compiler then produces an optimal scheduling at compile time. We also described how the existing OpenMP libraries can be used to implement concurrent execution of multiple CCU's.

References

1. S. Vassiliadis et al., The Molen polymorphic processor, *IEEE Transactions on Computers*, Vol 53, Issue II (Nov. 2004) ISSN: 0018–9340. pp. 1363–1375.
2. E. Moscu Panainte, K.L.M. Bertels, and S. Vassiliadis, Instruction Scheduling for Dynamic Hardware Configurations, *Proceedings of Design, Automation and Test in Europe 2005 (DATE 05)*, pp. 100–105, Munich, Germany, March 2005.

Chapter 11
Control of Dynamic Reconfiguration

Florian Thoma and Jürgen Becker

Abstract This Chapter describes the mechanisms used to control the dynamic reconfiguration aspects of the MORPEUS system. The base is formed by a real-time operating system and is topped by an allocation and scheduling system for reconfigurable operations.

Keywords Allocation • scheduling • real-time operating system

11.1 Dynamic Control Introduction

The dynamic control of the reconfiguration units is performed by the real-time operating system (RTOS) and the Predictive Configuration Manager (PCM), which is a HW-implemented unit supporting the RTOS and is described in more detail in Section 7.1.

The RTOS performs the following actions with information received from the PCM and the application/compiler:

- Allocation decision (choice of the implementation on the various reconfigurable units M2000, PACT, DREAM). If there are functionally equivalent implementations of the reconfigured operation it allows to make a choice at run-time depending on the platform/application status.
- Priority calculation of pending operations.
- Task execution status management.
- Resource request to the Predictive Configuration Manager for fine dynamic scheduling.

Information needed from the compiler is contained in the Configuration Call Graph (CCG). Other information used by the RTOS consists of:

F. Thoma (✉) and J. Becker
Universität Karlsruhe (TH), Kaiserstraβe 12, 76128 Karlsruhe, Germany
Florian.Thoma@itiv.uni-karlsruhe.de

N. Voros et al. (eds.), *Dynamic System Reconfiguration in Heterogeneous Platforms*, Lecture Notes in Electrical Engineering 40,

- Result of execution branching decisions
- Synchronisation points
- Task parameters

The configuration is handled by the RTOS and there is no direct communication between the application and the PCM. The communication between the application and the reconfigurable units is also only indirect and handled by the RTOS.

11.2 Real-Time Operating System

11.2.1 RTOS Overview

The real-time operating system is the interface between the application and the hardware; both dedicated and reconfigurable hardware. The application can use basic operating systems services as known from other operating systems. Additionally the RTOS offers the necessary services to support reconfigurable computing.

The configuration process involves several steps.

At first step, a SET system call will be used by the compiler to request the configuration of an operation on a reconfigurable unit.

This allocation is done dynamically by the RTOS during run-time using the alternative implementations of the operation, which are provided by the designer, on different reconfigurable units, depending on available resources, execution priority and configuration alternatives. The designer can also provide several implementations on the same heterogeneous reconfigurable engine, which are optimised for different criteria. The SET system call is non-blocking and returns immediately.

The second step is the execution of an operation. For the transfer of the parameters, the RTOS provides virtual exchange registers, which are mapped to physical register files inside the heterogeneous reconfigurable engines. After the filling of these registers by the compiler, the execute system call is used by the compiler to start the execution of an operation. The execute system call is non-blocking and returns immediately.

11.2.2 RTOS Features

The parallel running of operations which are mapped on the HREs with the code on the general purpose processor leads to synchronisation issues. Therefore the RTOS provides a BREAK system call which waits for the completion of the referenced operations. The compiler has to assure data consistency between parallel operations by using this system call. For sequential consistency, there is always a need for a break, corresponding to an EXECUTE system call before its data is further processed. It follows that the BREAK system call has to be a blocking operation which returns when all indicated operations are completed.

If an operation is no longer used, for example the calling loop has been left, the allocated resources have to be freed by the compiler with the RELEASE system call. The release system call is non-blocking and returns immediately.

The control of an operation is handled with the help of its status register. The RTOS writes commands like start execution or stop in this register. The operation is updating its status in the register whenever a state change occurs and indicates the change by an interrupt.

Application programming is supported by providing common operating system services including multithreading. The dynamic scheduler not only schedules the different threads but also computes the priorities of pending and near-future operations and schedules them for configuration and execution on the heterogeneous reconfigurable engines. The priorities are communicated to the Predictive Configuration Manager to allow the speculative prefetching of configurations. The Predictive Configuration Manager is commanded by the scheduler to configure the reconfigurable units. The allocation of the heterogeneous reconfigurable engine (HRE) to the different operations is closely linked to the scheduling to consider the configuration time/execution time trade-off if the preferred heterogeneous reconfigurable engine is already in use.

As it is preferable (for implementation and control reasons) that the heterogeneous reconfigurable engines do not perform memory access to the system memory, the operating system controls the DMA-controller of the system. This allows feeding the heterogeneous reconfigurable engines with data and transferring the result back to the system memory. The information about the memory arrays are provided by the application. At the same time, the RTOS ensures the data consistency by configuring the network on chip to use available buffers. The network on chip is then controlling the dataflow and the buffer switching on its own.

11.2.3 RTOS Structure

The RTOS has a layered structure which is shown in the Fig. 11.1. The bottom layer is the Hardware Abstraction Layer which provides a more uniformed access to the reconfigurable hardware and the system infrastructure. It provides virtual exchange register for the compiler which are mapped to the parameter registers in the heterogeneous reconfigurable engines. It also provides the basic for a pipeline service between the heterogeneous reconfigurable engines.

The middle layer is the Real-time Operating System Core. It provides the basic operating system services that are not related to dynamic reconfiguration and is based on an already existing eCos RTOS.

The top layer is formed by the dynamic reconfiguration framework called Intelligent Services for Reconfigurable Computing (ISRC). It provides the services for the configuration and execution of operations on the heterogeneous reconfigurable engines, e.g. the set, execute, break and release system calls.

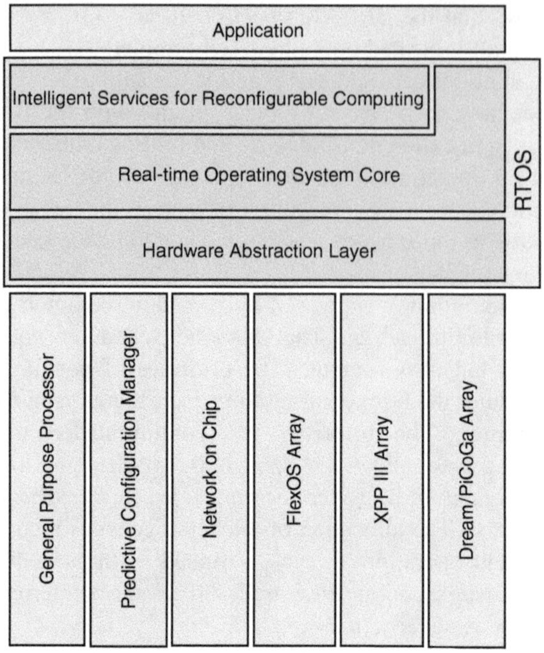

Fig. 11.1 Structure of the RTOS

11.2.4 RTOS Inputs and Outputs

Input for the RTOS consists of the configuration call graph (CCG), the MORPHEUS library and the implementation/configuration matrix. The MORPHEUS library contains the available operations, their implementations and the properties of these implementations like throughput, delay, size and power. The implementation/configuration matrix contains the list of the implementations used within a configuration and also indicates which configuration contains which implementation.

Output of the RTOS are the prefetching priorities and configuration commands for the configuration manager, execution commands for the operations on heterogeneous reconfigurable engines and control information for the network on chip and the DMA-controller.

11.2.5 eCos

eCos [1] is an open source, configurable, portable, and royalty-free embedded real-time operating system. An important focus of its development was the possibility to make the operating system highly configurable to achieve a good size/feature relation.

eCos was chosen for this project as the base for hardware abstraction layer and real-time operating system core as a compromise between the rich set of features of a linux kernel and the simplicity of minimalist kernels like TinyOS or μC/OS-II.

11.2.5.1 Features

- eCos is distributed under the GPL license with an exception which permits proprietary application code to be linked with eCos without it being forced to be released under the GPL. It is also royalty and buyout free.
- As an Open Source project, eCos is under constant improvement, with an active developer community, based around the eCos web site.
- Powerful GUI-based configuration system allowing both large and fine grained configuration of eCos. This allows the functionality of eCos to be customized to the exact requirements of the application.
- Full-featured, flexible, configurable, real-time embedded kernel. The kernel provides thread scheduling, synchronization, timer, and communication primitives. It handles hardware resources such as interrupts, exceptions, memory and caches.
- The Hardware Abstraction Layer (HAL) hides the specific features of each supported CPU and platform, so that the kernel and other run-time components can be implemented in a portable fashion.
- Support for μITRON and POSIX Application Programmer Interfaces (APIs). It also includes a fully featured, thread-safe ISO standard C library and math library.
- Support for a wide variety of devices including many serial devices, Ethernet controllers and FLASH memories. There is also support for PCMCIA, USB and PCI interconnects.
- A fully featured TCP/IP stack implementing IP, IPv6, ICMP, UDP and TCP over Ethernet. Support for SNMP, HTTP, TFTP and FTP are also present.
- The RedBoot ROM monitor is an application that uses the eCos HAL for portability. It provides serial and Ethernet based booting and debug services during development.
- Many components include test programs that validate the components behaviour. These can be used both to check that hardware is functioning correctly, and as examples of eCos usage.
- List source [2].

11.3 Intelligent Services for Reconfigurable Computing

11.3.1 Controlling HREs

The MORPHEUS SoC contains three vastly different reconfigurable units. The FlexEOS from M2000 is a fine-grained embedded FPGA. The DREAM from ARCES is a middle-granular reconfigurable unit with very fast context switches.

The XPP from PACT is coarse-granular array of processing elements optimised for pipelining algorithms.

With the difference in architecture comes a big difference in the configuration and execution control mechanisms. This has to be transparent for the user of design tools higher up the tool-flow. The difference of the configuration mechanisms is handled by a dedicated hardware called Predictive Configuration Manager. The services for reconfigurable computing sit on top of this to provide a uniform interface for the design tools.

11.3.2 Concept

The control of the HREs extends the SET/EXECUTE concept of the Molen compiler introduced in Chapter 10. The extensions have been necessary with regard to concurrent running threads competing for resources. In the original approach by Molen the instruction set of the processor is extended with special machine instructions. These instructions are replaced and extended by operating system calls.

- *SET:* the compiler notifies the operating system as soon as possible about the next pending operation to configure. The operating system uses this information to prepare configuration.
- *EXECUTE*: the compiler demands the execution of an operation.
- *RELEASE*: the configured operation is no longer needed and can be discarded.
- *BREAK*: wait for the completion of a operation for synchronisation.
- *MOVTX*: transfer data from the ARM processor to a specific exchange register of the HRE.
- *MOVFX*: transfer data from a specific exchange register of the HRE to the ARM processor.

11.3.3 Dynamic Scheduling and Allocation

The MORPHEUS platform is intended for many conceivable applications. The range goes from purely static data stream processing to highly dynamic reactive control systems. These control systems react on changes of the environment like user interaction, requested quality of service, radio signal strength and battery level in mobile applications. These changes can significantly change the execution path of the application and the priority of the various operations. Static scheduling and allocation is not sufficient for such reactive systems.

The topics of scheduling and allocation are tightly related on reconfigurable systems. For this reason the RTOS provides a combined dynamic scheduler and allocator. It determines during run-time schedule and allocation of the operations requested within one thread and in parallel threads. The combined scheduler/

allocator determines which alternative implementations of the requested operations are available and their associated costs. The usage of configuration call graphs (see Section 11.3.4) provides knowledge about the structure of the application. All this information is used to update the schedule of pending operations with the goal to improve metrics like overall throughput and latency.

When reasonable the scheduler moves operations from an over-loaded heterogeneous reconfigurable engine to another heterogeneous reconfigurable engine between consecutive calls. A requirement for the usefulness of dynamic allocation is that the application designer offers a choice of implementations for as many operations as possible.

The user has a choice between various scheduling and allocation strategies.

11.3.3.1 First Come First Serve

This is the most basic scheduling methodology. The only criterion for scheduling is the sequence of requests. When a task is finished the next operation from the queue is examined if the needed HRE is available, otherwise the scheduler waits till the needed HRE is freed. When the operation is available for several HREs they are all tried in sequence of falling throughput.

11.3.3.2 First Come First Free

Instead of one global queue this methodology uses one queue per HRE. At request time the operation is added to each HRE queue where it is available. As soon as a HRE is finished it is used by the next operation in its queue and the operation is removed from all other queues. This method maximises the overall capacity utilisation of each HRE but can result in frequent reconfigurations.

11.3.3.3 Shortest Configuration First

The delay for configuration depends on two factors. First one is the size of the configuration bitstream. It also depends on transfer speed of the bitstream to the HRE which again depends on the used memory of the system. The off-chip sram and flash memories are significantly slower than the on-chip configuration sram managed by the Predictive Configuration Manager. The schedule requests the prefetching state from the PCM and calculates a configuration time. An available HRE is then configured for the operation with the shortest configuration time. This methodology minimises idle times due to preferring recurring and short operations.

11.3.3.4 Maximum Throughput First

This methodology uses the operation with the highest throughput waiting for a specific HRE. This maximises overall data throughput but can harm responsiveness.

11.3.3.5 Minimal Delay First

This methodology uses the operation with the lowest delay waiting for a specific HRE. This maximises responsiveness but can harm data throughput.

11.3.3.6 Weighted Score

All the above methods for scheduling and allocation focus on one criterion while completely ignoring the influence of the other factors for the performance of the system. With the exception of First Come First Serve and First Come First Free they also suffer from the chance that operations starve and never get processed. The solution used here is to compute a score for each operation by weighted sum of all criteria including a waiting-time. The formula for the score can be computed as

$$P = K_C \sum_{i=0}^{3} \frac{ConfigurationSpeed_i}{BitstreamSize_i} + K_T Throughput + K_D \frac{1}{Delay} + K_W WaitingTime$$

The weighting factors K_C, K_T, K_D and K_W are adjustable and allow tailoring of the scheduling to the specific needs of an application. The capital-sigma sum is necessary to consider the different characteristics of the different levels of the configuration memory hierarchy. The index 0 refers to the configuration exchange buffer (CEB) and 1 refers to the on chip configuration memory. The external configuration memory is divided between sram (index 2) and flash memory (index 3). The equation can be easily extended for other storages as hard disk drives or network storage. The values of BitstreamSize_0 and BitstreamSize_1 depend on the current prefetching state and can be determined by polling the PCM. Throughput and delay are two dimensions of performance measurement. WaitingTime is increased every time an operation is not scheduled to execute and is used to prevent starvation.

11.3.4 Configuration Call Graphs

The dynamic allocation is improved by using foresight of coming operations. The RTOS uses for this purpose the configuration call graphs provided by the Molen compiler.

The compiler provides one CCG per thread. It contains the sequence, including branches and loops, of the operations and their configuration. During run-time the scheduler traces the running of the different application threads through their configuration call graphs to have a global estimate about which heterogeneous reconfigurable engine is going to be available in the near future for use and the next pending operations. This probability information is communicated to the Predictive

Configuration Manager which uses it for prefetching configuration bitstreams from external memory to the on-chip configuration memory which results in a significant reduction of reconfiguration overhead. The PCM feeds back information about the prefetching state of the bitstreams to the RTOS. The scheduler uses the prefetching state for allocation decisions, e.g. it can favour a slower implementation which is already in the on-chip configuration memory against a faster implementation which at first has to be loaded from external memory.

11.3.5 DMA/DNA

The DMA and DNA provide communication mechanism between various HREs and memory and between heterogeneous reconfigurable engines for data transfers.

The dynamic allocation of operations can make it necessary to change the source or destination address of data transfers which make it essential to handle the communication by the operating system. It provides an API with a unified interface for transferring data with the NoC or the DMA for application programmers or upstream tools like SPEAR. The linking of a transfer to an operation allows to migrate the transfer automatically to the new HRE.

For more details on the data communications infrastructure see Chapter 8.

11.4 Conclusions

In this chapter we introduced the real-time operating system as the base layer of the dynamic control of the MORPHEUS system. We explained how the system calls for controlling the HREs relate to the Molen compiler. Finally we presented a selection of algorithms available for dynamic scheduling and allocation and showed the advantages of the Weighted Score algorithm over other algorithms.

References

1. eCos Reference Manual, http://ecos.sourceware.org/docs-latest/ref/ecos-ref.html.
2. eCos User Guide, http://ecos.sourceware.org/docs-latest/user-guide/ecos-user-guide.html.

Chapter 12
Specification Tools for Spatial Design

Front-Ends for High Level Synthesis of Accelerated Operations

Arnaud Grasset, Richard Taylor, Graham Stephen, Joachim Knäblein, and Axel Schneider

Abstract Reconfigurable architectures like the MORPHEUS platform represent a significant breakthrough in the embedded systems research. The heterogeneity and the complexity of the reconfigurable units integrated in such architecture is however an issue to develop applications. This chapter presents how few specification tools improve the development of applications on reconfigurable units. These tools are used as front-ends for synthesis from high-level models.

Keywords ADeVA • Array-OL • Cascade • communication • loop transformation • MADEO • model of computation • refinement • SPEAR • SpecEdit • tiling

12.1 Introduction

The ambition of the MORPHEUS reconfigurable platform is to deliver processing power competitive with state-of-the-art System-On-Chip, while maintaining flexibility to a broad spectrum of application, and user-friendliness. An innovative aspect of the MORPHEUS approach resides in the close coupling of a development tool chain (see Chapter 9) with a dynamically reconfigurable architecture (see Chapter 3). The productivity of application development on the platform is

A. Grasset (✉)
Thales Research & Technology, France
arnaud.grasset@thalesgroup.com

R. Taylor and G. Stephen
CriticalBlue Ltd, United Kingdom

J. Knäblein and A. Schneider
Alcatel-Lucent, Nuremberg, Germany

N. Voros et al. (eds.), *Dynamic System Reconfiguration in Heterogeneous Platforms*, Lecture Notes in Electrical Engineering 40,

improved thanks to a seamless design flow from a high level description toward target executable code.

A MORPHEUS application is described by a C program, executed on a main processor, that includes some calls towards "operations" (as explained in the Chapter 10). The execution of these "operations" is accelerated on several Heterogeneous Reconfigurable Engines (HRE), detailed in Chapters 4, 5 and 6. We can simply define an operation as a part of an application (i.e. a function), implemented on one HRE, manipulating and processing complex data structures (e.g. image frames) called macro-operands. The calls mentioned above consist in configuring the HRE and then launching the execution of the operation. An HRE is (re)configured by loading a bitstream in a configuration memory integrated in the HRE.

An ambition of the MORPHEUS toolset is to abstract the heterogeneity and the complexity of the architecture in such a way that software designers are able to program it (without knowledge and experience of HREs). As a manually built library of pre-defined operations could not be exhaustive, the Spatial Design framework of the MORPHEUS toolset is dedicated to the design on the various reconfigurable units of the "operation" mentioned above. It includes programming the HREs, generating configuration bitstreams but also managing the communications to feed these accelerators with data.

To this end, the goal of Spatial Design is twofold. The first objective is to hide the HREs heterogeneity and abstract the hardware details for the programmer. A Control Data Flow Graph (CDFG) format is used as an intermediate and technology independent format inside the framework. High-level synthesis techniques are used in a tool, called MADEO, that acts as a back-end for the framework (Chapter 13).

A second objective is to provide domain-specific models and languages to application programmers. In this way, operations can be specified at a high-level of abstraction (improving design time and flexibility), without sacrificing performances. An innovative and varied panel of specification and system-level tools is at the designer's disposal to describe accelerated operations using:

- A graphical formalism with the type of boxes found in Simulink [1], combined with C language descriptions for the boxes behavior. This model is particularly well suited to signal processing applications. Data-parallelism mapping techniques, implemented in a tool named SPEAR, can take benefit from the inherent parallelism of operations. The C language description of the boxes being compiled into an ARM executable, the Cascade tool automatically extracts a CDFG for use by downstream synthesis tools.
- A formal specification captured in the SpecEdit tool. SpecEdit helps to enter specifications in a structural way and supports their refinement towards implementation and verification in many ways. It is possible to create CDFG for MADEO from SpecEdit data.

In this chapter, we first present the SPEAR tool and its graphical formalism. Second section shows how the Cascade tool can generate a CDFG from a C description. Then, Section 12.4 gives an overview over the specification framework SpecEdit.

12.2 The Spear Design Environment

12.2.1 Mapping Operations on Reconfigurable Engines

12.2.1.1 Challenges of the Mapping

Reconfigurable architectures represent a new option to design embedded systems, that have well known specific requirements in terms of performance and power consumption. However, programming and exploiting these architectures is a challenge for application developers, that puts a brake on a wide deployment of reconfigurable technologies.

Mapping an application on MORPHEUS architecture starts by identifying some kernels or critical parts in the application to offload from the processor, and selecting the most appropriate HRE to accelerate them. Such a task is a challenge that usually requires some designer's know-how to take the best decisions. All the more, programming the HREs involves several languages and programming models. The designer also has to cope with the limited memory space, an eventual impact of communication costs on performance speed-up and the hardware/software heterogeneity.

As a result, mapping operations on HREs is time consuming and error prone. Therefore, it limits the design space exploration and makes application design costs prominent.

12.2.1.2 A Tool for Mapping Applications on Architectures

A quick feedback on implementations is the key for a fast design space exploration from a "golden model". Edward Lee [2] has stressed the need of *Models of Computation* and dedicated tools to program heterogeneous architectures. SPEAR is a tool developed at Thales Research & Technology, for the mapping of data streaming applications on heterogeneous multiprocessor architecture. It is based on a *Model of Computation* suitable for signal and image processing applications, and more precisely for data and computation intensive applications composed of nested-loops. It follows the Y-chart philosophy [3] where architecture and application are fully independent models for application portability and for anticipating future application and needs. It is in line with the MORPHEUS approach that targets both high-performance and flexibility needs.

In the MORPHEUS design tool flow, SPEAR is a key tool at the interface of the high-level synthesis of operations and of the SW compilation flow; enabling their cooperation in an integrated and coherent toolset (Fig. 12.1). The SW compilation flow and the RTOS manage the creation of a software executable running on the main processor, from the application code (see Chapters 10 and 11). This executable controls the reconfiguration and the execution of HREs but also programs DMA transfers. The communications are however managed in the Spatial Design

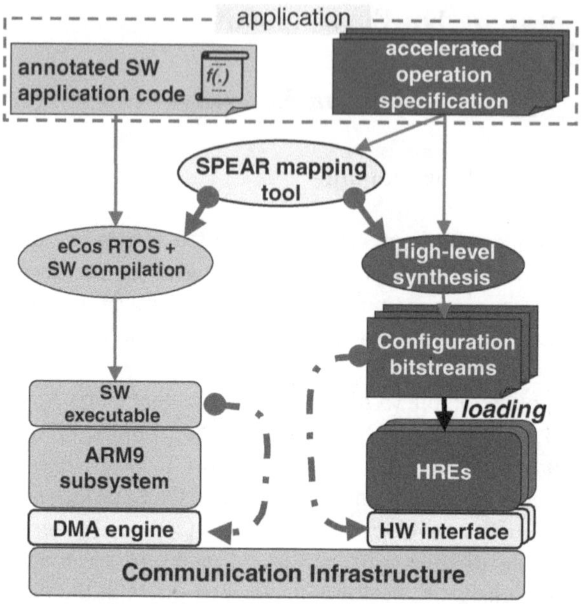

Fig. 12.1 Overview of SPEAR contribution to MORPHEUS toolset

framework. The high level synthesis produces a configuration bitstream from an operation specification.

SPEAR helps to capture a specification of an operation accelerated on a HRE, that is independent of the targeted HRE. It then enables to manage in a coherent framework both the HW interface of the HRE (i.e. local buffers and addressing mechanism) and the data transfers. Connections of SPEAR with high-level synthesis and SW compilation tools avoid risks of incoherencies between the layout of data in local buffers and main memories, the DMA transfers, the addressing mechanisms of the HREs to access to local buffers and the operation specification.

12.2.1.3 Contributions of SPEAR to the Toolset

SPEAR helps to map an operation on an HRE in the following ways:

1. Its graphical environment offers to the user an intuitive method for loops transformation (e.g. loop fusion, tiling) in order to optimize the memory footprint of the mapped operation.
2. SPEAR automatically generates some communication parameters to feed the HRE. They are forwarded to the SW compilation flow, contributing in this way to a seamless hardware/software integration.
3. SPEAR is able to do a functional simulation for an early validation of the operation specification, as well as to generate "operation-specific" address generators for high-level synthesis and rapid-prototyping.

12.2.2 *Specification of Operations*

12.2.2.1 The Array-OL Formalism

The Array-OL formalism [4] is well suited to represent deterministic, data intensive and data-flow applications such as the kind of operations accelerated on HREs. Operations are modeled in this formalism as a directed acyclic graph, whose nodes represent tasks exchanging multi-dimensional data arrays (Fig. 12.2a).

A task executes a set of perfectly nested loops, whose kernel is a basic function. This function, called elementary function, is written in a subset of the ANSI C programming language. In this way, Array-OL combines the flexibility and the legacy of the C language with the advantages of a capture in a building-block environment (user-friendliness, block reuse, close to signal processing algorithm models [1]). The loop bounds are constant values. The Array-OL formalism puts in evidence the linear access patterns of a task within the arrays of data that they produce and consume. The parameters of elementary functions are fixed size arrays. A function can take as input parameters only a subset of input arrays. In the same way, an output parameter can be assigned to only a subset of the output array. This subset can vary at each iteration. It corresponds to an atomic data pattern, extracted from the

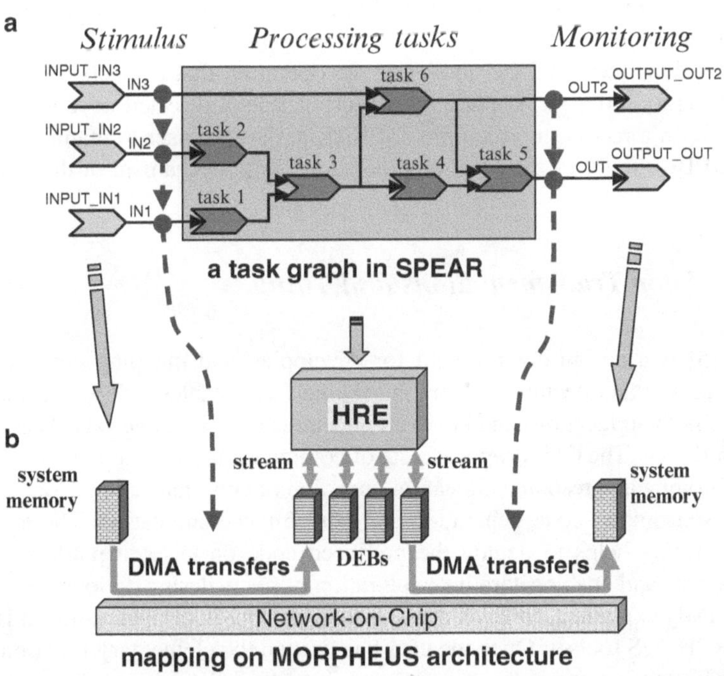

Fig. 12.2 Overview of an operation mapping on a HRE

array at regularly spaced positions. The formalism assumes a single assignment of data in output arrays. So, there are no dependencies between loop iterations.

12.2.2.2 Application Mapping on the MORPHEUS Architecture

In agreement with the MORPHEUS programming model, the input data of an operation are initially stored in a system memory (on-chip or off-chip memory) as well as the results of the operation at the end of the HRE processing (Fig. 12.2b). But an HRE can only access to a set of local memories attached to each reconfigurable engines, called Data Exchange Buffers (DEB). Therefore, data structures are transferred from the system memories to the DEBs, processed by the HRE and the results are transferred from DEBs to system memories. The data transfers on a Network-on-Chip and the HRE processing are pipelined in order to avoid stalls of the HREs and reduce latency by overlapping processing time and communication time. While an HRE processes a chunk of data, the next chunk of data is transferred. A way to manage this pipeline is to use "ping-pong" buffers. The DEBs are managed in a swapping way so that when a DEB is used by the HRE, another is used for the data transfers.

The processing in the reconfigurable units is stream-based. Some kind of addressing mechanism enables to feed the HRE with data streams by reading data in DEBs. The processing inside the HRE is also pipelined, using pairs of DEBs as intermediate buffers between stages of the pipeline. The MADEO synthesis considers that the swapping of all couples of DEBs is synchronous and happens during the pipeline progress.

SPEAR generates a CDFG model of the operation that respects this execution model. Operation design is made by capturing it as a directed acyclic graph and mapping it on a reconfigurable units. SPEAR assists the user to manage layout of data in DEBs, the DMA transfers and the addressing mechanism of the HRE.

12.2.3 Loop Transformations and Tiling

SPEAR [5] is a design environment for developing and mapping applications on multi-processor architectures. A graphical interface enables to capture high-level models of both applications and target architectures in an intuitive way. The mapping is human-driven. The GUI proposes a set of commands to the user, covering application partitioning and resource allocation, insertion of communications, loop transformations, scheduling, code generation and performance simulation. The automated code generation helps to deploy the produced code on targeted platforms, which reduces errors and enables iterative exploration of many design options.

As already explained, SPEAR deals with the mapping of operations on HREs in the MORPHEUS toolset. Once the user has written the elementary functions in the C programming language, he just captures the Array-OL schema of an operation. In order to accommodate the data size of an operation with DEBs sizes, well known loops transformations [6] and tiling techniques are then at the disposal of the user in the SPEAR framework to optimize data locality of the operation. The following loops transformations are relevant in the context of MORPHEUS project:

- *Loop strip-mining*: decompose a loop in an equivalent nest of two loops
- *Loops interchange*: change loop nesting order
- *Loops fusion*: replace two loops with identical fixed bounds by one loop whose kernel is the concatenation of the two previous kernels

In Fig. 12.3, a picture of 128 × 128 pixels is sent to an operation doing a binarisation of each pixel. Assuming a coding with 4 bytes by pixel, a full picture requires 65.5 Kbytes and cannot fit in a DEB of 4 Kbytes.

Figure 12.4 shows the previous example after loops strip-mining and interchange. To implement the operation into the architecture, the picture is decomposed into 16 sub-pictures of 32×32 pixels. The operation is then executed in several processing phases consuming and producing sub-pictures. Each loop of 128 iterations is strip-mined in 2 nested loops of 4 and 32 iterations (4×32 = 128). The loops of 4 iterations are interchanged to be outermost loops. Iterations of the outermost loops are then scheduled in different processing phases during the mapping of the operation. Now, one sub-picture only requires a memory space of 4Kbytes and can fit in one DEB.

A sub-picture is in a more general way called a chunk of data. SPEAR helps the user to decompose an operation execution in many processing phases; so that each phase operates on a chunk of data. A chunk of data contains required data to feed all

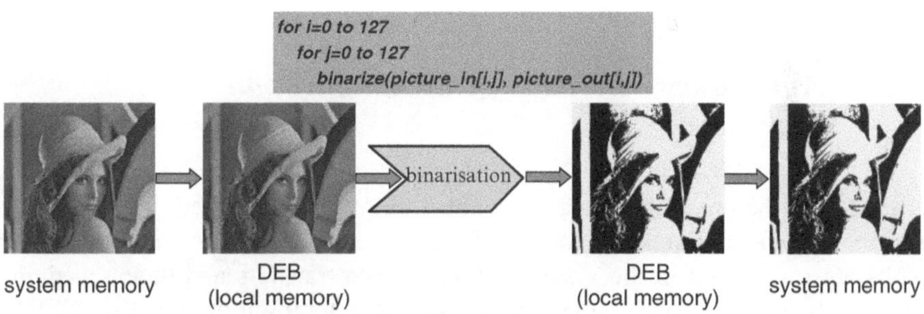

Fig. 12.3 Example of mapping without tiling

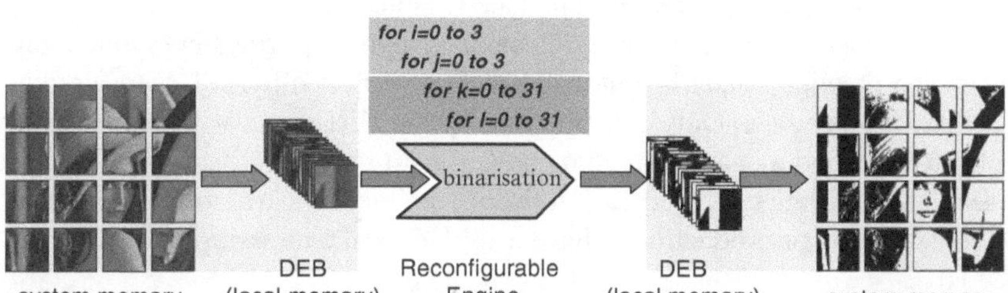

Fig. 12.4 Illustration of the tiling technique

loop iterations scheduled inside a processing phase. So, the designer can reduce the memory footprint and defines chunks of data by applying loop transformations.

In case of operations composed of multiple tasks, intermediate arrays between tasks can also be stored in DEBs. Intermediate memory requirements are reduced in the same way by loop transformations. The pipelined execution model presented in Section 12.2.2.2 involves that all the tasks composing the operations are scheduled with a same number of processing phases. To schedule loops iterations in many processing phases, the corresponding loops have to be common to all the tasks and a fusion of these loops is done. In this way, a chunk of data produced by a task contains all data required by the following tasks. If necessary, the user can split and permute the loops in order to enable the loop fusion.

The loop transformations also influence the scheduling of loop iterations inside a phase in an intuitive and implicit way. The need of intermediate data storage can sometimes be avoided by a careful scheduling. Exchanges of data arrays between tasks are refined into data streams when implementing the operation on the HRE. The data order in the produced and consumed streams is fixed by the scheduling of loop iterations inside a phase. If a produced stream cannot directly be consumed, data are buffered in local memories. Intermediate buffering is for instance necessary when data of an array are produced in a row-major ordering, but consumed in a column-major ordering. It is equivalent to a corner-turn in traditional software processing.

When the user has applied loops transformations to build a mapping strategy, code generation feature of SPEAR tool is then used to automatically generate the communication parameters.

12.2.4 Automatic Generation of Communication Parameters

As explained before, input data of an operation are transferred from a system memory to the Data Exchange Buffers (DEB) to feed the HRE. After computation, results are transferred in the system memory. Data transfers and their scheduling depends on both the operation and on the way of mapping it on the HRE. SPEAR automatically generates few communication configuration parameters, used by the software compiler to generate some operation-specific code programming the Direct Network Access (DNA) module (see Chapter 8).

We assume a row-major ordering for the data layout of arrays in system memory (C language practice). Data layout in DEBs is determined by SPEAR according to the defined data chunks. Only data required by the HRE processing are transferred. SPEAR tries to find a data layout that enables burst transfers, assuming if necessary sparse reading of data by the HRE. If it is not possible, data are stored in the order of their consumption or production to have a simple local addressing law implemented in the HRE.

We consider only chained multi-block transfers, programmed in the DNA through linked lists of block descriptors. Transfer of an array is composed of a set of block transfers. A block is a set of data contiguous both in the source and destination memories. All the blocks have the same size, but they can have different source addresses and destination addresses. So, data blocks stored at non-contiguous addresses in system memory (regular and sparse location of blocks in memories) can be stored at contiguous addresses in DEBs. The generation of block descriptors is simply based on nested-loops. Base addresses of blocks are processed by incrementing addresses in nested loops at each end of a loop iteration. Base addresses of a block on the source and destination sides follow thus a multidimensional linear equation:

$$block\_address_{i_0,\ldots,i_{nb\_loops}} = base\_address + \sum_{j=0}^{nb\_loops} i_j * step_j \tag{1}$$

In Eq. (1), i_j is the index of block for the jth loop and $step_j$ the address step for the *j*th loop. The assumptions made by the Array-OL formalism guarantee that such a solution is always possible. The communication parameters are generated considering use of DEBs with a ping-pong scenario.

12.2.5 A Stream-Based Processing

One elementary function is implemented by one stream-based processing block (CDFG of this block being compiled by Cascade tool). As a result, data are streamed in and from these blocks, assuming a one-to-one relation between an array and a stream. As data are not streamed in or from the HRE but stored in DEBs, SPEAR generates or configures an interface between these blocks and the DEBs. If data streamed from a block can be streamed directly in the destination block, no glue is necessary for the communication between these blocks. Otherwise some glue with intermediate storage is automatically inserted to connect these blocks (Fig. 12.5).

Three kinds of blocks implement the glue between the stream-based implementations of the elementary functions and the DEBs:

- *Input controllers*: An input controller produces a data stream from a data structure stored in two DEBs operating in a ping-pong scheme. It is a kind of *n*-dimension address generator programmed to read a *n*-dimension pattern of data in DEB.
- *Output controllers*: On the contrary of an input controller, an output controller writes a data stream into a Data Exchange Buffer. The written data structure is then transferred through the Network-on-Chip in a system memory.
- *Connectors*: A connector is a component transforming a data stream into another stream by rescheduling, duplication and removal of data. It is composed of two address generators: one for writing a stream into a DEB, and another for reading data in the DEB with a different ordering during the next computation phase.

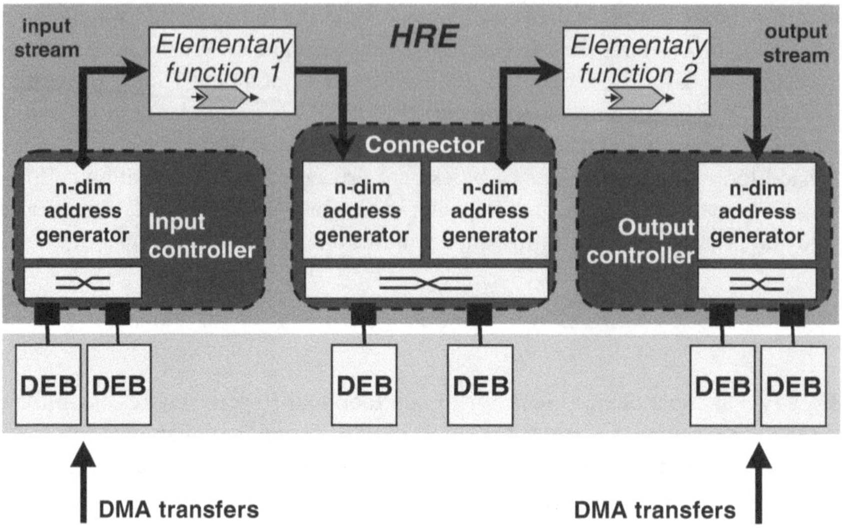

Fig. 12.5 Implementation of an operation with connectors and IO controllers

The address generators produce a finite sequence of memory addresses according to a multi-dimensional linear equation. Pattern overlapping and data skipping is possible. A linear address pattern is defined for a phase and thus repeated in a cyclic way for each processing phase. These components are specific for each operation, and depends on the scheduling done with the method explained in Section 12.2.3.

As said before, the processing inside the HRE is made in a pipelined way. The blocks inserted by SPEAR are the bounds of each stages, and so need a couple of DEBs operating in a ping-pong fashion. During the synthesis done by MADEO, a global controller will be synthesized to synchronize the swapping of DEBs. This controller synchronizes with the ARM or the DNA using a handshake mechanism (Chapter 8), so that the pipeline progresses when data to process are available and when DEB space is available to store results. As a synchronization is done for each block transfer, SPEAR determines a number of block transfers between each swapping.

The SPEAR framework features automatic code generation at multiple levels of abstraction:

- A high-level model in SystemC to verify by simulation the operation capture, as well as the communications and the address generators parameters.
- Code of connectors, input and output controllers under the form of Control Data Flow Graphs (CDFGs). From this intermediate level description, the synthesis with MADEO tool implements these components either by programming hardwired address generators inside the HREs, or by implementing application-specific address generators in reconfigurable logic of the HRE.

- Makefiles and synthesis scripts to drive other modules and tools. From the SPEAR environment the user can generate CDFGs of elementary functions thanks to Cascade tool, connect them with the SPEAR's CDFG and do the synthesis with MADEO.
- A testbench in Verilog in case of mapping on the M2000 HRE. Stimuli and results of the SystemC simulation are then used to verify the netlist obtained after synthesis by MADEO tool.

12.3 Cascade Design Flow

12.3.1 Introduction

CriticalBlue's Cascade solution allows software functionality implemented on an existing main CPU to be migrated onto an automatically optimized and generated loosely coupled coprocessor. This is realized as an automated design flow that provides a bridge from a standard embedded software implementation onto a soft core coprocessor described in RTL. Multiple coprocessors may be attached to the main CPU, thus architecting a multi-processor system from a software-orientated abstraction level.

The CriticalBlue CASCADE tool analyzes ARM processor software binary and enables the automatic generation of an optimized programmable coprocessor. The resultant coprocessor is generic and can be targeted to any ASIC or FPGA target.

The generated coprocessors are not fixed function hardware accelerators, but programmable ASIPs. Their combination of functional units, internal memory architecture and instruction set are derived automatically by analysis of the application software; as such they represent a powerful instrument for system partitioning. Migration of functionality onto coprocessors does not impose the loss of flexibility implied by the existing manual processes of developing hardwired RTL accelerator blocks.

The primary usage of such coprocessors is for offloading computationally-intensive algorithms from the main processor, freeing up processing resources for other purposes. Clearly there will be a limited set of blocks within a typical SoC where a custom hardware implementation remains the only viable option, perhaps due to physical interfacing requirements or extreme processing needs beyond that practically achievable with programmable IP. It should be noted that the Cascade flow includes the capability for customer IP to be utilized within the generated coprocessor RTL, thus widening the applicability of the approach.

Fundamentally, the advantage provided by this design trajectory is about exploitation of parallelism. The parallel execution of the main processor alongside coprocessor(s) provides macro level parallelism whilst the microarchitecture of the coprocessors themselves aggressively exploits Instruction Level Parallelism (ILP). In this sense the goals of the flow are identical to those of the manual process of offloading functionality onto a hardwired block.

Since this process is heavily automated the user can rapidly explore many different possible architectures. Being a fully programmable machine, a coprocessor can be targeted with additional functionality beyond computationally-intensive kernels with little additional area overhead. This allows design exploration unencumbered by typical concerns of RTL development time and over-reliance on inflexible hardwired implementations.

One unique aspect of the Cascade solution is that it implements directly from an executable binary targeted at the main processor, rather than from the C implementation of an algorithm. A static binary translation is performed from the host instructions into a very different application-specific instruction set that is created for each coprocessor. We have found that this flow provides a significant usability enhancement that eases the adoption of this technology within typical SoC projects. Since existing software development tools and practices can still be used the existing environments and knowledge can be leveraged to full effect. Moreover, the offloaded functionality remains in a readily usable form for use on the main CPU, allowing greater flexibility in its deployment for different product variants. Figure 12.6 provides a flow diagram of the steps in using Cascade from a user perspective. The initial steps for targeting the flow are identical to those of a fully manual approach.

12.3.1.1 Cascade in the MORPHEUS Context

In the MORPHEUS project the Cascade tool was used to generate MADEO CDFGs for the elementary functions. The C code representing the elementary functions is compiled into ARM binary code and then the front end of Cascade is used to analyze that binary code and produce an abstract CDFG representation of the data and control flow of the function. This may then be synthesized into gates by the MADEO tool.

Cascade is used to extract a high level Control and Data Flow (CDFG-HLL) model of a SPEAR sub process. The model is an individual description of one computation element that can be composed with other similar models into a larger spatially mapped dataflow system with the use of automated connectivity synthesis between the processes.

Cascade interfaces with the Step/Express Java API provided by UBO to produce a library of process CDFG-HLLs that can then be read by the MADEO+ tool in order to supply technology mappings for the supported reconfigurable fabrics.

The front end analysis stage of Cascade enables a compiled ARM executable to be analyzed and a CDFG-HLL generated for use by MADEO+. The CDFG describes all of the operation level control and data flow dependencies within the SPEAR sub process. Since Cascade reads an ARM executable rather than a C source file, an ARM compiler must be included in this flow. The output CDFG-HLL is

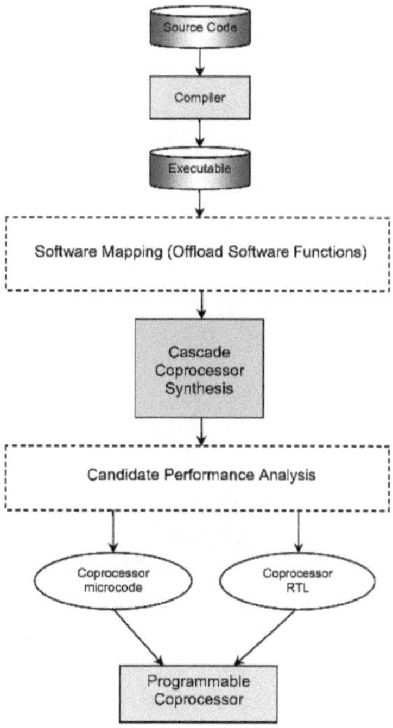

Fig. 12.6 Cascade design flow

constructed using the Step/Express API for Java, allowing interoperability of the various design tools associated with the spatial mapping flow. The generated CDFG is untimed and technology independent. The physical mapping to a CDFG-LL is performed using the MADEO+ tool flow.

The Cascade design flow for use as part of a MORPHEUS deployment is shown in Fig. 12.7. User interaction with Cascade is via a command line interface, such that the Cascade portion of the flow may be easily scripted. The first parameter of the call to the Cascade command line interface is the name of the ARM executable that contains the SPEAR sub processes to be translated into CDFG-HLLs. The remaining parameters (of which there should be at least one) are the names of the functions that represent the SPEAR sub processes that are to be modeled.

The following pages describe the stages of binary analysis, graph construction and transformation and stream analysis required in the Cascade portion of the MORPHEUS flow.

12.3.1.2 ELF/Symbol Analysis

This step reads the ARM executable file in ELF format. Analysis is performed on the layout of the code in memory and the symbol table is read to determine the relationship between symbols and software functions or data variables. Analysis is also

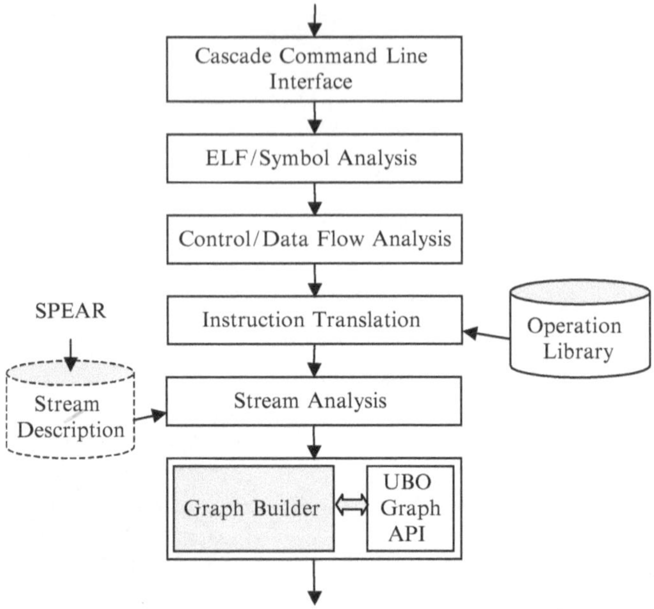

Fig. 12.7 Cascade design flow within MORPHEUS

performed to determine whether or not each instruction influences control flow, so as to provide a basis for subsequent control and data flow analysis.

12.3.1.3 Control Data Flow Analysis

Control flow analysis of the code reveals the control flow relationships between basic blocks (control flow graph nodes) in the code from which high level control flow relationships can be inferred. Data flow analysis determines the transference of data via registers in the application.

A very basic representation of control flow is constructed using the instruction knowledge acquired during Elf/Symbol Analysis. This knowledge is also used to represent all the data flow within the basic block. At this point, further graph transforms have still to be applied to infer more C-like control constructs and also to remove many data flow operations that are redundant in the MORPHEUS context.

12.3.1.4 Instruction Translation

This step examines the input ARM instruction stream and partitions each individual instruction into a sequence of primitive operations and data moves between those operations. In the context of MORPHEUS these primitive operations are defined by the primitives available in the operation library. If certain intrinsic functions are supported as primitives then this step transforms the C function calls into primitive mappings.

12.3.1.5 Stream Analysis

The purpose of this step is to recognize accesses being performed on the input/output parameters (i.e. streams) of the process and rewrite these as receive/send primitives in the Step/Express Java API. These primitives communicate with the communication connectors associated with the process and either read data from and input or write data to an output. The stream analysis also provides a certain amount of validation to ensure that the ordering in which the parameters are accessed corresponds to the stream description so that these reads/writes can be activated in the hardware using a basic FIFO hardware mechanism. This validation is limited somewhat by the fact that all analysis on the binary is static.

Much of the information required for each stream cannot be determined by static analysis and must therefore provided to Cascade in a sub function description file. SPEAR produces such a file, which contains information on:

- Sub function name
- Parameter index in the sub function signature
- Parameter direction (input/output)
- Parameter data type
- Parameter dimension size and index

The following example can be used to illustrate the sub function description file format. Consider the function:

*void func1(int x[50][100], int y[50][100], int *output)*

The corresponding sub function description file would be:

```
<?xml version="1.0" encoding="UTF-8"?>
<subFunctionContainer>
<subFunction name="func1">
<streamParameter name="x" index="0" direction="input" dataType="int" numDim="2">
<paramDimension size="50" dim="0" />
<paramDimension size="100" dim="1" />
</streamParameter>
<streamParameter name="y" index="1" direction="input" dataType="int" numDim="2">
<paramDimension size="50" dim="0" />
<paramDimension size="100" dim="1" />
</streamParameter>
```

<streamParametername="output" index="2" direction="output" dataType="int" numDim="1">
<paramDimension size="1" dim="0" />
</streamParameter>
</subFunction>
</subFunctionContainer>

The sub function description file essentially describes one or more C function signatures, except for the return type, which is implicitly void in the MORPHEUS flow. Cascade uses the function signature information in the sub function description file to separate data operations pertaining to the stream from those which may be common in ARM code, but not necessarily relevant in the MORPHEUS context. Such an operation could be a register preserve–restore operation, which stores in memory the value of a live-in, non-parameter register at function entry and restores the register value at function exit, allow the register to be used for general data storage during function execution.

12.3.1.6 Hierarchical CDFG Construction

Constructing the Hierarchical CDFG (HFG) involves using graph transformations, control flow recognition heuristics and stream information acquired from the Stream Analysis stage to refine the basic CDFG that was created in the Control Data Flow Analysis stage.

More control flow analysis is performed to identify and reconstruct the C-like aspects of the ARM instructions. The Control flow hierarchy identified in Control Data Flow Analysis is transformed into conditional loops and if/else statements. The transformed control structure, consisting of the conditional loops, if/else statements and other C-like control constructs, is then in a format suitable for processing by MADEO.

The data flow is optimized to eliminate unnecessary register reads and writes within a control flow block. In an ideal situation, this should leave a set of reads of the registers that are live into the block and a set of registers whose values are live out of the block. Intermediate operations within the block would ideally consist of only data flow operations. However, it should be noted that elimination of all intermediate register accesses may not always be possible due to a variety of reasons, including the maintenance of a large set of local variables.

Within each basic block, the data flow is further optimized using the Stream Analysis information. This information is used to determine which operations in the data flow graph pertain to the ARM's own address arithmetic and register preserve–restore operations (and in which case may be carefully removed by Cascade) and those which apply to the input and output data.

With both the data flow and control flow graphs in a condition ready for translation, the HFG is transformed into the MADEO+ format using the UBO Step/Express API. The generated CDFG provides a hierarchical view of control and data flow within the SPEAR sub process. At the lowest level of the hierarchy, data flow is represented by operations consisting of nodes that correspond to the basic arithmetic, logical and storage operations in the assembly code of the original executable. Control flow is represented by hierarchical nodes that correspond to the C-like constructs of if/else statements, conditional loops and fixed iteration loops.

12.3.2 *Example: Simple Finite Impulse Filters*

The remaining subsections in this section show a simple example of CDFG generation using Cascade in the MORPHEUS context. Section 12.3.2.1 shows the C source input and a sub function description file describing the two functions to be translated. Compilation of the C source file into ARM.axf file format is required to prior to supplying the.axf file and the sub function description file to Cascade. Section 12.3.2.2 displays the output of each function translation in graphical format.

12.3.2.1 Inputs: C Code and Sub Function Description File

C Code for Simple Filter Functions FIR_HI and FIR_LOW

```
#include <stdio.h>
void FIR_HI(int pixel[5], int *outputpixel) {
   *outputpixel = pixel[0] * −1 + pixel[1] * 2 + pixel[2] * −1;
}
void FIR_LOW(int pixel[5], int *outputpixel) {
      *outputpixel = pixel[0] * −1 + pixel[1] * 2 + pixel[2] *
6 + pixel[3] * 2 + pixel[4] * −1;
}
```

Sub Function Description File

```
<?xml version=“1.0” encoding=“UTF-8”?>
<subFunctionContainer>
<subFunction name=“FIR_HI”>

<streamParameter name=“pixel” index=“0” direction=“input”
dataType=“int” numDim=“1”>
<paramDimension size=“5” dim=“0” />
</streamParameter>
<streamParameter name=“outputpixel” index=“1”
direction=“output” dataType=“int” numDim=“1”>
<paramDimension size=“1” dim=“0” />
</streamParameter>
</subFunction>
<subFunction name=“FIR_LOW”>
<streamParameter name=“pixel” index=“0” direction=“input”
dataType=“int” numDim=“1”>
<paramDimension size=“5” dim=“0” />
```

</streamParameter>
<streamParameter name="outputpixel" index="1" direction="output" dataType="int" numDim="1">
<paramDimension size="1" dim="0" />
</streamParameter>
</subFunction>
</subFunctionContainer>

12.3.2.2 Outputs: Function Graphs

The shaded square nodes (Receive Nodes) at the top of Fig. 12.8 correspond to the reception of input streams. At the bottom of each graph, the shaded square node (Send Node) sends the output stream. In the middle, the nodes correspond to the dataflow part of the FIR.

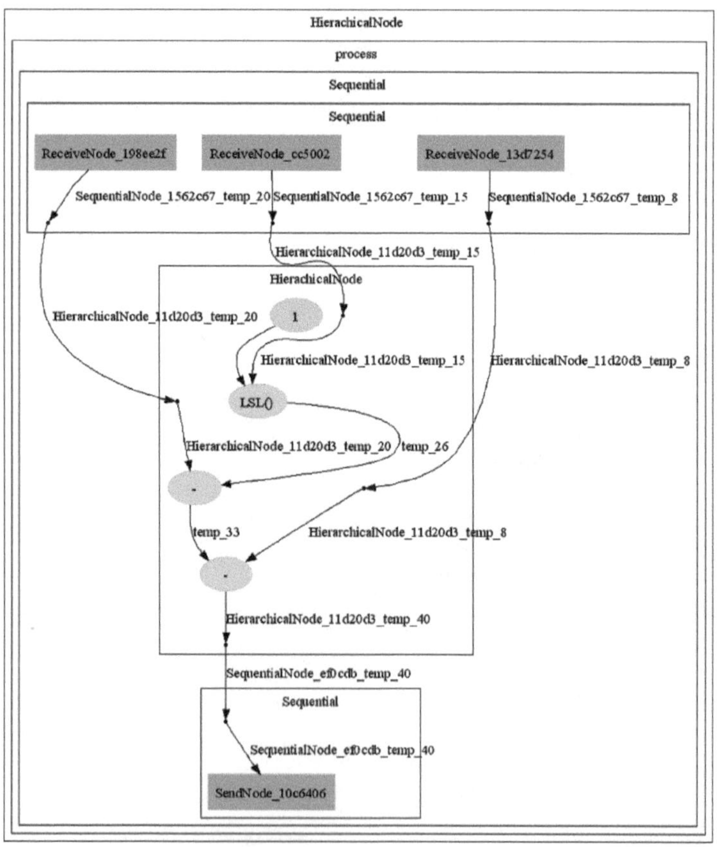

Fig. 12.8 The graph of the translated FIR_HI

12.4 SpecEdit

This section describes a prototype tool which has been developed at Alcatel-Lucent in order to facilitate the system specification work and to reduce the error rate of this task. Latter aspect was motivated by the fact that the earlier a problem is found, the less expensive its solution is. The name of the tool is **'SpecEdit'**.

12.4.1 Introduction to SpecEdit

From the SpecEdit perspective a specification is a collection of system requirements. Requirements can be structured in requirement groups. Such groups again may contain other groups. This can be compared to a directory tree, where the directories correspond to the requirement groups and the files are similar to the requirements. Each of the objects described can have a number of content of different kinds like e.g. text, graphics, formal specification data or timing information.

It is possible to add assistants for definition of such requirement content using an open plug-in approach. Such an assistant supports the user in the entry of particular specification information and checks the data for consistency. For a physical hand out of a specification document such an assistant provides a print driver which defines the document representation of its specification data.

Figure 12.9 depicts the GUI of SpecEdit for illustration. On the left side the requirement navigator is located. This browser shows the requirement structure which reflects the chapter hierarchy of the generated document.

At the upper edge of the right part of the GUI object relevant information is entered. Below that the assistant region is situated. In this screenshot the text definition assistant is shown. Currently this is an embedded OpenOffice instance, but other word processing tools are conceivable. All features of such a word processor like e.g. graphics or diagrams are available for the specifying person in order to describe the intention of the requirement.

Below the assistant region a table of defined names and mnemonics is presented. Typically in system specifications a lot of (sometimes confusing and cryptic) short names are used. The table here supports the user with name completion capabilities in order to reduce the risk of misspelling and wrong usage.

12.4.1.1 ADeVA

Another important assistant is the one which provides support for ADeVA [7], a method for formal specification which is based on a tabular format. ADeVA is a means to specify control flows in a way similar to the usage of finite state machines known especially in hardware design. Admittedly not every specification aspect can be defined using this approach, but it enables to describe the most critical aspects,

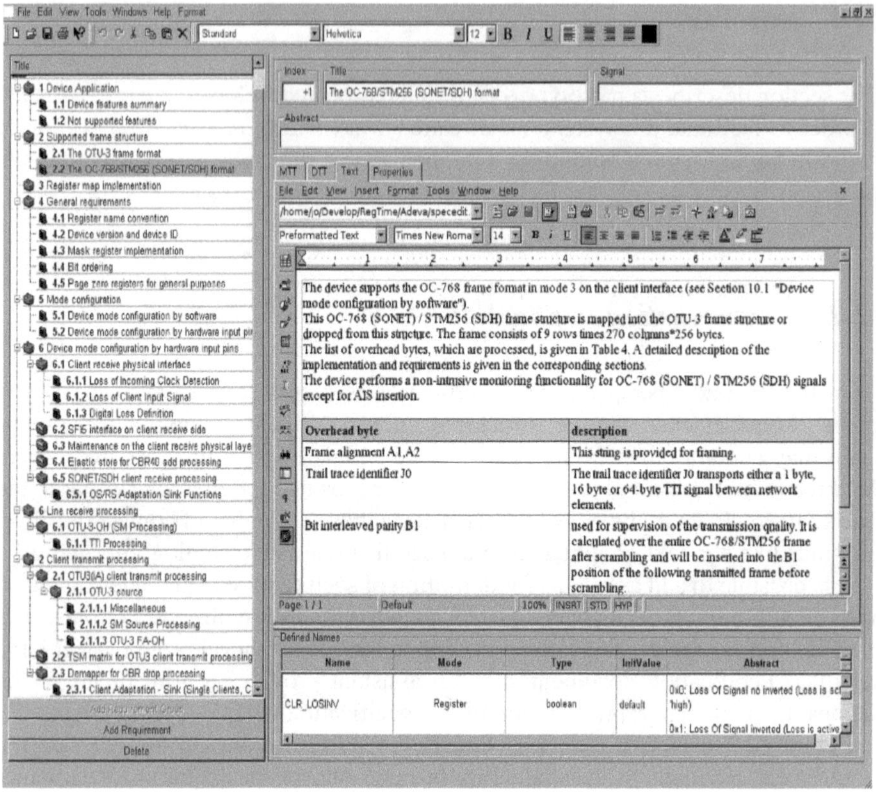

Fig. 12.9 The GUI of SpecEdit

i.e. the aspects which typically are the most error prone and the most development time consuming ones. Basically an ADeVA description is a tabular representation of state transitions of the system, from which a model is generated. Model languages, which are supported in the current version, are: VHDL (most elaborated), C, System-C, SMV and SAL. The latter two are languages which are proprietary to model checker tools of the same name. This model can be used for simulation or formal verification on specification level fostering to the motto of Spec Edit: “Find errors as earlier as possible”. Under a different perspective ADeVA can be regarded as a simple way to define an executable specification, which can be verified with dedicated tools. For the formal verification job properties are developed which prove that the specification really meets the intention of the requirements. These properties can be reused during the entire design process in order to guarantee that the implementation complies with the specification. It is obvious that the properties which are appropriate for the specification level will not fit other design stages. Thus, refinements of the properties have to be performed. This aspect is discussed Section 12.4.3.

Figure 12.10 gives an impression how SpecEdit supports ADeVA and what such an ADeVA table could look like.

In this figure the ADeVA assistant is focused instead of the text assistant. An ADeVA table as displayed defines the setting of a system state in dependency on external and internal signals. This kind of definition is called MTT (mode transition table). The first column determines the start state and the second column the target state. A transition is performed if the condition which is determined when all the conditions in the following columns become true. The '@' character in front of a condition definition means, that this condition is only true when there has been a change on the inputs in a way that the condition evaluates to true/false.

Every requirement may come with such a table. The tables communicate with signals of a particular type. This type is defined in the defined names list. The resulting model contains a number of asynchronously cooperating automatons, where e.g. a state transition in one automaton may trigger a series of other state transitions in other automatons.

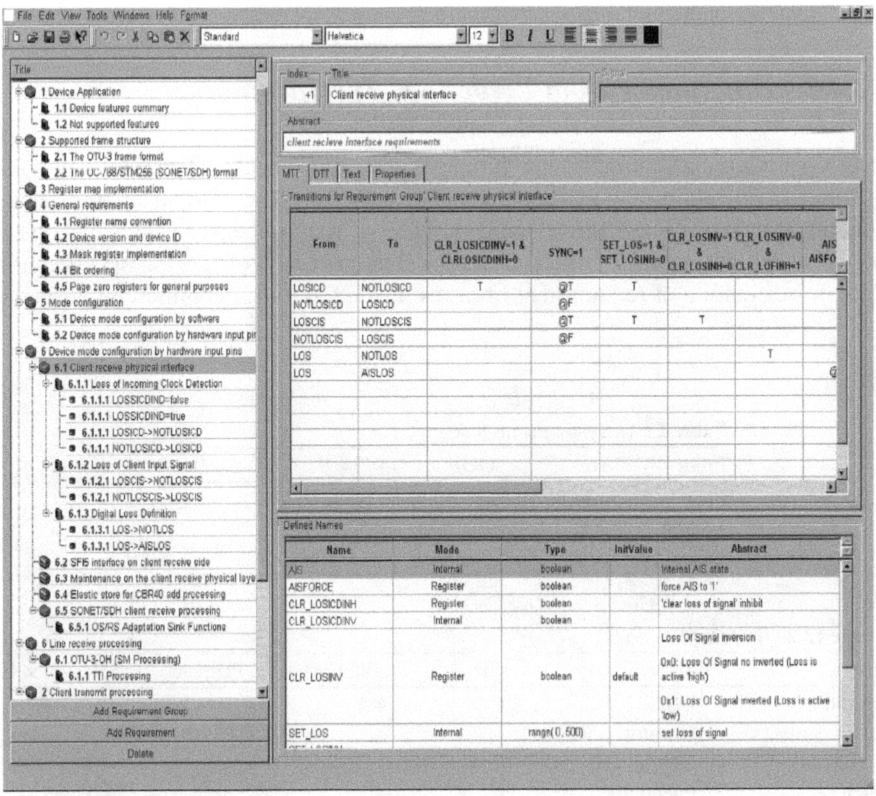

Fig. 12.10 ADeVA support in SpecEdit

12.4.1.2 Example for Interconnected MTTs

Table 12.1 exemplary shows a MTT definition. In this MTT a system state 'SYS1' is defined which has the values: first, middle, last. The system state 'SYS1' changes from first to middle if: 'OOF' is false, 'IN2' is equal to 0 × 0 and 'IN1' becomes unequal to 9. The other rows can be read in the analogue way. 'OOF', 'IN1' and 'IN2' could be system inputs or internal signals being listed in the 'Defined Names Table'.

Table 12.2 shows a second MTT. In this MTT a system state 'SYS2' is defined which has the values: monitor, latch and unequipped. The system state changes from 'monitor' to 'latch' if: 'AIS' is false, 'IN2' is not greater than 0 × 10 and 'SYS1' becomes unequal to last. 'IN2' is a system input and 'SYS1' is an output of the other table. The rows following describe transitions in an analogue way.

In addition to MTTs so called DTTs (Data Transition Tables) describe the output behavior of the system, i.e. they define the value of an output signal in dependency on the values of input- and internal signals.

System functionality is decomposed in a collection of MTTs and DTTs from which a model for simulation-based or formal verification is generated.

12.4.2 *Inheritance from Predefined Specifications*

It is very common that particular functions (= requirements) of previous designs are integrated into a new system under development. The amount of transferred functionality may vary from a small number of requirements to all the requirements which belong to a large sub-design. The next step of reusing specification data is the creation of requirement libraries. In such libraries common requirement

Table 12.1 System state SYS1

From	To	OOF	IN1! = 9	IN2 = 0 × 0
First	middle	F	@T	F
Middle	last	T	@T	–
Last	first	F	–	@T

Table 12.2 System state SYS2

From	To	AIS	SYS1 = last	CPU read	IN2 > 0 × 10
Monitor	Latch	F	@T	–	F
latch	monitor	–	–	@T	–
monitor	unequipped	F	–	–	@T
latch	unequipped	F	–	F	@T
unequipped	monitor	F	–	–	@T

(groups) are stored together with all their content like descriptive documentation or ADeVA tables.

Against this background a very powerful referencing and reuse concept is required. The reference concept of SpecEdit is based upon the idea of inheritance. Instead of a requirement or group a reference to an object (i.e. requirement or requirement group) of another specification can be added to the requirement hierarchy. Object content can be overwritten or mapped on an elementary basis. Such an element could be the type of a specific variable or the name of a MTT signal. Overwriting means, the value of this element is entirely replaced by a new value. Mapping, however, means that the original value (i.e. the value of the underlying requirement) is modified according to defined rules (e.g. search & replace).

12.4.3 Refinement of System Specifications Towards Implementation

It has already been said, that SpecEdit is a tool which is targeted to accompany the design process from system specification to implementation and finally system verification.

In this context two aspects are of special interest:

1. Refinement of specification information, primarily ADeVA tables for implementation purposes
2. Refinement of specification properties for verification purposes

The way SpecEdit supports the refinement process itself is also based upon the idea of inheritance as described in Section 12.4.2. A refinement level of the specification inherits everything from the previous one, i.e. initially all the requirements are transferred directly. The refinement step itself is done requirement by requirement e.g. by redefinition of types and ADeVA tables. In parallel the verification properties must be adapted to the refined version as well.

12.4.3.1 Refinement of Specification Information for Implementation Purposes

Most profit is gained by the reuse of proven specification data. In principal the reuse of specification data is achieved by refinement, i.e. the abstract description is enhanced by more and more implementation specific facets. Three areas for refinement have been identified:

1. Introduction of timing constraints
   As a first step towards implementation the design must be partitioned in terms of hardware/software parts, clock domains and more. For this design space

exploration job timing constraints are introduced in order to be able to check the functionality of the specification under different timing circumstances. In particular this means that specific function e.g. state transitions are enriched with timing information.

2. Reorganization of the requirements
   At this stage the requirements are structured according to logical specification aspects. These aspects are not always inline with the implementation architecture. Thus, when moving from specification level to the implementation view the hierarchy of requirements may be reorganized to reflect the implementation structure.
3. Increasing accuracy of variable types and introduction of clocks in MTTs
   This item of the list is related to the final target that the refined tables can be synthesized by conventional tools in particular the FPGA synthesis tool MADEO.

12.4.3.2 Refinement of Specification Properties for Verification Purposes

A strong means for the important and tedious system verification task is the ability to reuse the properties which have been set up during the specification process. Such properties describe the behavior of a system on an abstract level and are therefore appropriate to check system behavior over different levels of abstraction. Of course the properties also have to be reworked from level to level, but the amount of effort is small compared to the effort which has to be spent on implementation level. Thus the probability to introduce errors in the refinement process is smaller than on implementation level. Additionally if implementation and verification refinement steps are done by different teams (cross-checking), it is unlikely, that both teams make the same mistakes.

12.4.3.3 Positioning SpecEdit in the MORPHEUS Design Flow

Figure 12.11 shows the role which SpecEdit plays in the MORPHEUS tool flow. SpecEdit models are most suitable for being mapped to the fine grain eFPGA core FlexEOS, since the interlocking of the SpecEdit models with the RTOS especially RTOS scheduler and software in general seems difficult. Thus, SpecEdit is a somewhat isolated solution for specification and implementation of functions to be integrated in the eFPGA HRE of the MORPHEUS chip.

12.5 Conclusions

Three tools have been presented in this chapter to produce Control Data Flow Graphs (CDFGs) from an operation specification.

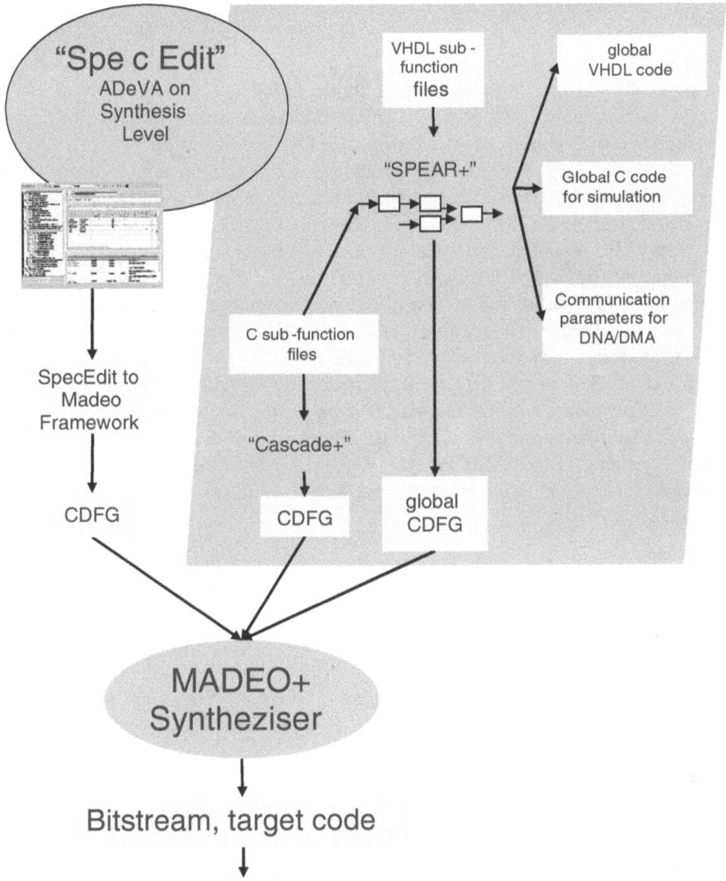

Fig. 12.11 Integration of SpecEdit in the MORPHEUS design flow

SPEAR enables to capture operation as a graph of tasks. Loop transformations are applied to cope with the limited size of the local memories. Then, SPEAR generates communication parameters as well as some kind of address generators. Cascade generates CDFGs from the C description of the elementary functions associated to each task.

Spec Edit is a tool that facilitates the system specification work. A specification can be progressively refined toward implementation and verification. An operation can be specified in this framework.

The generated CDFG are used as an intermediate model in the Spatial Design framework, between the tools presented in this chapter and the MADEO tool. Next chapter shows how MADEO is used for the synthesis of operations on the HREs.

References

1. The Mathworks, Simulink® 7 Getting Started Guide, 2008.
2. E. Lee and A. Sangiovanni-Vincentelli, A framework for comparing models of computation, *IEEE Transactions on Computer-Aided Design*, 17(12), 1998.
3. K. Keutzer, S. Malik, A. R. Newton, J. M. Rabaey, and A. Sangiovanni-Vincentelli, System-level design: orthogonalization of concerns and platform-based design, *IEEE Transactions on Computer-Aided Design*, 19(12), 1523–1543, December 2000.
4. P. Boulet, Array-OL revisited, multidimensional intensive signal processing specification. Research Report RR-6113, INRIA, February 2007.
5. E. Lenormand and G. Edelin, An industrial perspective: pragmatic high-end signal processing environment at Thales, *Proceedings of the 3rd International Workshop on Synthesis, Architectures, Modeling and Simulation*, SAMOS, 2003.
6. M. E. Wolf and M. S. Lam, A data locality optimizing algorithm, *Proceedings of the ACM SIGPLAN 1991 Conference on Programming Language Design and Implementation*, PLDI '91.
7. W. Haas, U. Heinkel, and S. Gossens, Behavioural Specification for Advanced Design and Verification of ASICs (ADeVA), *GI/ITG/GMM-Workshop Methoden und Beschreibungssprachen zur Modellierung und Verifikation von Schaltungen und Systemen Tübingen*, 2002.

Chapter 13
Spatial Design

High Level Synthesis

Loic Lagadec, Damien Picard, and Bernard Pottier

Abstract MORPHEUS accelerators programs are process compositions working on local memories. They produce instruction level parallelism and concurrent memory accesses. Spatial design is a middleware between high level compilers and circuits mapped on the accelerators. Its core is a model for process code used by high level development tools and for synthesis on heterogeneous targets. The framework also ensures system performance by overlapping communications and computations.

Keywords Control Data Flow Graph • synthesis • concurrent processes • Madeo

13.1 Spatial Flow: Transport Specification, CDFG and Synthesis to Reconfigurable Target

13.1.1 Spatial Design Presentation

Spatial design is defined as the activity that moves data from memories to memories, and through the processing circuits set-up on accelerators. It is better to consider this activity as a whole for several reasons described in this introduction. Let us assume that a reconfigurable accelerator process data in a loop involving main memory read, transport to the accelerator, processing, transport from the accelerator, and write back to high level data structure in main memory (Fig. 13.1). Some aspects of this loop follow.

First, accessing a group of data spread over memory generally requires address computation usually handled by a processor before effective access to memory. A common example is to retrieve a record in a multidimensional array, and then a particular field in this record. As our accelerators are distant from

L. Lagadec (✉), D. Picard, and B. Pottier
Université de Bretagne Occidentale, France
loic.lagadec@univ-brest.fr

N. Voros et al. (eds.), *Dynamic System Reconfiguration in Heterogeneous Platforms*, Lecture Notes in Electrical Engineering 40,

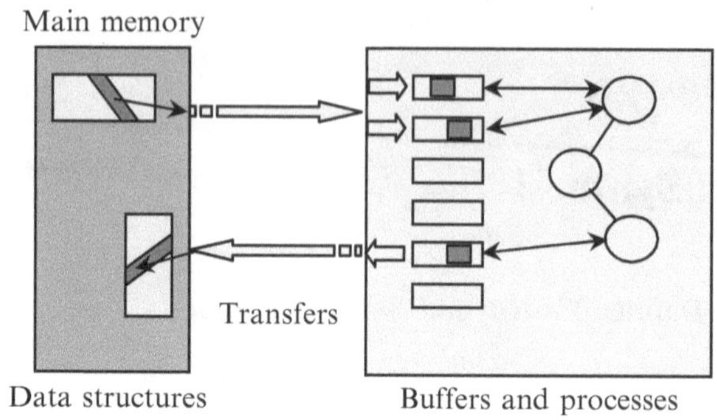

Fig. 13.1 Data are moved from high level data structures in memory to local buffers connected to the processing circuits

main storage, it is necessary to handle address computation, in order to produce streams of data.

A second aspect related to data spatial distribution is *where* the data go in local memories. The choice of these locations is obviously important for system activities and the computation structure. Computing processes on the reconfigurable part are organized to read and write local memory locations, in harmony with the transport system.

A third aspect of local data distribution is the bandwidth of the process group to local memories. When very intensive computations are necessary, a possible choice is to configure process circuits enabling several concurrent memory transactions rather than sequential accesses to one memory.

The computation itself is probably not a single process, but a group of processes working together on local memories with in-circuit data exchanges. Furthermore, several accelerators can be involved to support computations from a group of processes, with communications appearing between processes. Thus, *Spatial design* also includes the problem of the decomposition of a computation into a network of cooperating processes.

The first step in modeling an execution loop is to represent the process level organization including system level communications. Then, it becomes necessary to consider the process computation structure, in terms of numeric operations, control operations, and memory accesses conforming to the data flow. Synthesis for heterogeneous targets will handle this computation structure and produce adequate reconfigurable circuits or back-end specifications for proprietary tools bound to these targets.

Before going into details, it can be noticed that the computing loop over reconfigurable devices must be synchronized externally with the main program and with the operating system in charge of resource allocations. The data flowing to the accelerator must be scheduled to appear in buffers at the adequate moment, to be extracted soon enough, and processes themselves have further constraints not to overrun exchange registers.

This section begins by explanations on the structure and sequencing of the compute loop, then it describes the abstraction used for computing kernel to be synthesized on reconfigurable circuits.

13.1.2 Transporting Data

Within the MORPHEUS project, reading and writing data in memory is achieved by the *SPEAR* methodology that also covers local accesses based on the *connectors* concept. Behind this implementation there are software mechanisms and architectural support needing some explanations.

MORPHEUS definition for the reconfigurable accelerator behavior is that data must be collected from main memory, and result written back to main memory. In other words, the accelerator works as an autonomous entity called by the application program according to *Molen* protocol.

The accelerator activity can be quite long with intensive access operations on memory, so it is necessary to support this activity in hardware independently from the processor threaded operations. In fact, address generation is achieved by *Direct Memory Access* controllers (DMA). DMAs are programmed by the operating system as it is typically a place where race conditions can occur. Considering the main processor and at least one accelerator, we can see that we are in a shared memory situation where several actors can pump and store data concurrently.

The normal way to program DMA is to set-up command blocks before starting a transfer activity. These blocks are chained or aggregated together into a queue. The content of a DMA block defines source and destination for data, as memory access, or peripheral ports. In our case, these locations can be associated to data structures in memory, and accelerator wrapper ports giving access to local memories.

Addressing activity by the DMA is based on the definition of a starting pointer, the width of a data element, the offset to the next element, and the count of elements. This addressing method is known as a linear addressing mechanism and is associated to many software paradigms usual in the world of parallelism. An example is *Parallel Virtual Machine* (PVM) pack and unpack functions that allow building or reading communication buffers from linear data distributions. The addressing capability of the DMA includes fetching and storing of row, columns, diagonals, sub-blocks of arbitrary sizes within multi-dimensional data structures. It also applies well to multimedia data, as found in elementary stream structures.

Thus, data access is close to usual packing operations and could probably be covered by commonly used software libraries (Fig. 13.2).

13.1.3 Handling Data in the Accelerators

Data streams produced by a DMA appear at an accelerator port to be stored in local memories. Many possibilities can be considered to store these data, but they all

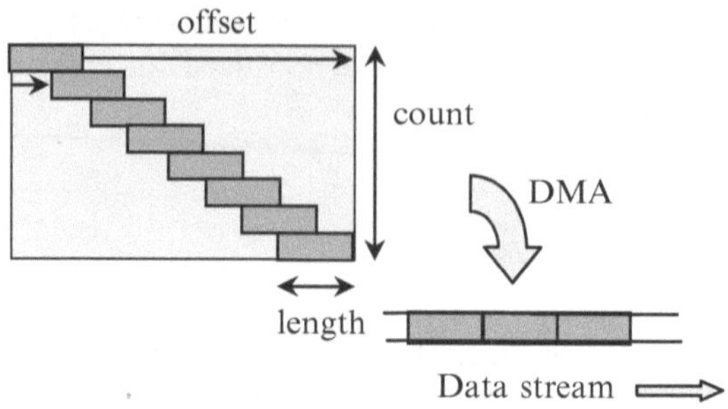

Fig. 13.2 High level data structures being addressed as linear transactions from a DMA

need to be considered together with local processing behavior. Basically the first thing to consider is the number of addressing mechanisms on local memories.

In the case of a single port, each memory is allocated to the communication or the processing activity as shown in Fig. 13.3. We refer to this technique as the ping-pong scheme, where address ports are alternatively connected to the communication port or the reconfigurable engine. Referring to the figure, during one phase, buffers B2 and B3 are used to compute result, and at the following phase it will be the turn of B1 and B4. The control of the accelerator guarantees that data will not be wasted as it will be shown.

A noticeable point is that this architecture avoids copying data. Buffers are simply switched using a multiplexer connected to a port in the configured process.

To get acquainted of further questions concerning the memory management, the implementation choice in Spatial Design was chosen to be simple and efficient. Not all the opportunities have been explored.

As an example, spreading data over several input memories allow concurrent accesses with possible impact on performances (Fig. 13.4). This is feasible, provided that local memories are organized in a single address space, and a local DMA can control spreading over these memories. Another interesting example is the use of a local memory access network that provides a shared memory access for configured processes [2].

13.1.4 Sequencing Execution

The sequencing issue is also appearing at system level with the necessity to coordinate operations involving one or several accelerators. Several options are possible, but it is mandatory to observe several constraints. One obvious constraint is data dependency: (1) do not start a computation before needed data are available locally, and (2) do not send back result to main memory before computation is finished.

Another constraint comes from buffers or memory space available locally. If a small amount of buffers are available, it is needed to manage their use

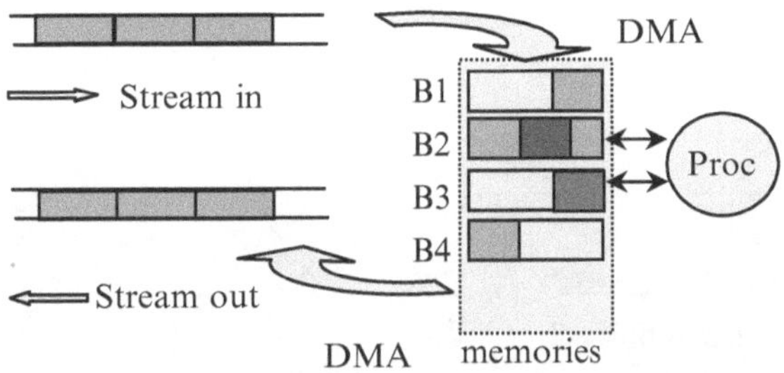

Fig. 13.3 Local data distribution: B1 is being filled while B4 is dumped in a data stream. Process Proc is using B2 and B3 as random memory processing data from B2 to produce results in B3

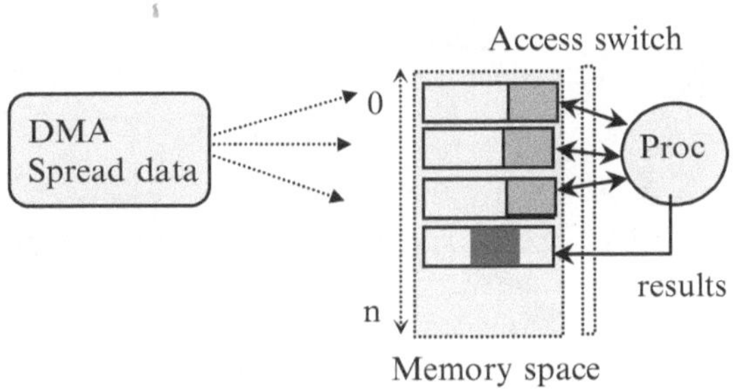

Fig. 13.4 Reducing memory bottleneck by spreading data over several local buffers, with concurrent memory access for a compute process with large bandwidth to memory

conservatively, recycling used buffers to receive data and filled buffers to send back data to main memory.

In terms of performances, it is important to overlap communication and operations as much as possible in order to have the maximum bandwidth for the compute circuit.

To reach these objectives *Spatial Design* proposes to consider communication and computations as concurrent processes organized as a system pipeline. The first stage of this pipeline is the place where incoming communications take place. The second stage is for transforming data into local buffers. The third stage is for outgoing communications, back to main storage, or sending to other accelerators (Fig. 13.5). Dependencies are managed on the basis of synchronization barriers between all processes involved in the three stages. Communications can effectively be processed concurrently by DMAs and it is safe to consider that they are occurring in this way provided that they are described by independent DMA command blocks.

Phases are thus defined to be (1) a group of independent incoming communication tasks, (2) a group of processing tasks associated to processes mapped on the reconfigurable device, and (3) a group of outgoing communication tasks. To allow the pipeline to work, it is necessary to set-up a minimal local controller in charge of barrier synchronizations, and once a barrier is reached, in charge of accelerator management. A noticeable need is to make decision on which buffer to allocate to which

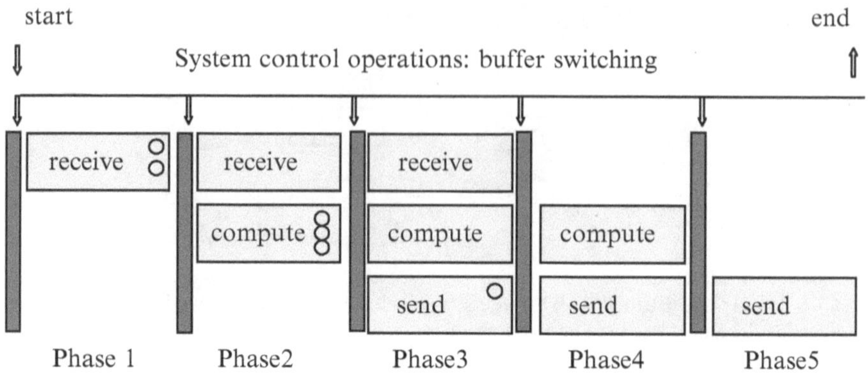

Fig. 13.5 Macro pipeline for feeding, computing and dispatching results on an accelerator. Bullets figure subtask appearing in communications and computations

task for the next phases. Practically, the controller must receive termination signals from all the DMAs involved in communication task and from all processes involved in communications. It also has the responsibility to deliver start signals to DMAs and processes, to activate next pipeline phases. The controller is activated by the operating system, and delivers end of computation signal once all data are processed.

13.1.5 Communication Structure

As explained above, communications are defined by DMA command blocks. These blocks associate linear addresses in memory and destinations. Usually several such communications need to be produced for each pipeline phases. A simple example is a matrix vector product where part of lines and columns must be sent in a particular order to allow processing to take place, plus periodically a write back of the resulting vector (Fig. 13.6).

Thus several transfer blocks are generally set-up for each phase defining communication tasks. These blocks are written in memory by the high level compiler or tools such as SPEAR, as a preamble of the accelerator computation.

13.1.6 Computation Structure

Computation is organized as group of processes started synchronously at each phase beginning. In principle, these processes can access local memories randomly with the restriction that memories need to be allocated to communication or computation processes. Each memory has only one access mechanism. In the SPEAR implementation, this memory access is given to one *connector* process having the

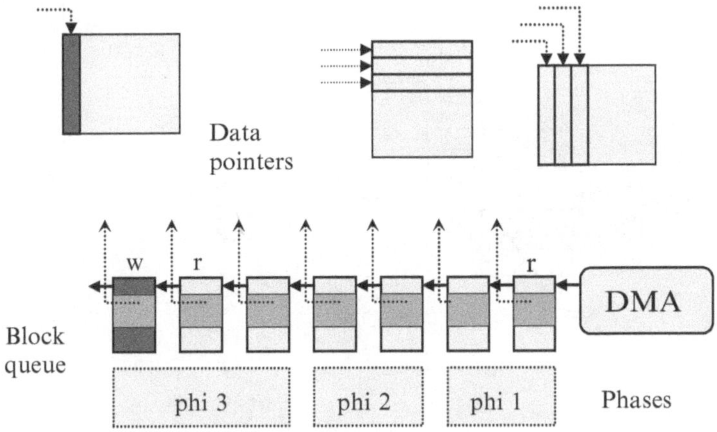

Fig. 13.6 DMA block queue structure showing three phases, with communication sub-tasks. The grey command block is for a write back operation and other ones for sending row and columns to a compute engine

responsibility to dispatch or collect data from compute process. Figure 13.7 shows a process organization with such connectors, compute processes, and controller process enabling buffer switches using multiplexers.

Conceptually, the group of processes implements a computation flow using local memories as temporary locations for buffering or data sharing and reorganization. Processes can exchange data based on blocking read and write operations operated on local channels. These operations are used to assemble memory access processes and computation processes.

Group of processes are started by the local controller and signal their termination to this controller, contributing to the accelerator pipeline barriers. It will be explained later how synthesis tools take care of the blocking channel operations.

13.1.7 Algorithm Abstraction: CDFG

We have explained that global data moves are based on system level transfers executed by DMA operations. Local data moves are executed by processes running on heterogeneous reconfigurable circuits.

It is now time to consider how it is possible to address *heterogeneous reconfigurable engine* (HRE) as found on the MORPHEUS SoC. The problem is new and interesting since the entry specification for these targets are very different in nature. Fine grain reconfigurable devices are usually addressed using hardware description language (HDL), such as VHDL, or Verilog, while coarse grain targets prefer syntaxes close to sequential programming such as C derivatives.

An initial objective in MORPHEUS was to produce a *single programming flow* for these targets, allowing future HRE to be implemented and used easily by

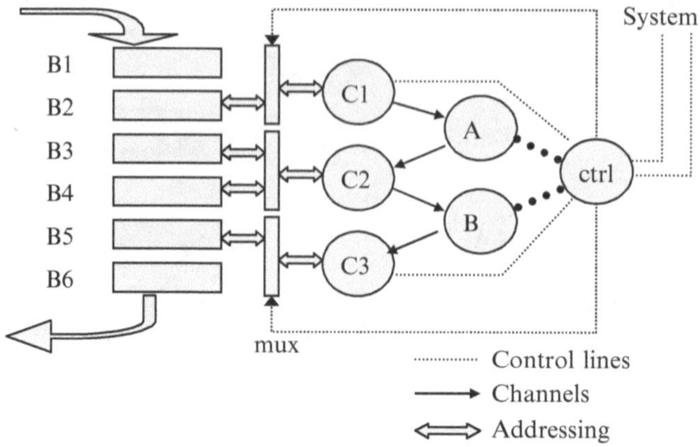

Fig. 13.7 Local memory and compute processes working with a local controller. Ci are memory access processes, A and B are compute process, while Ctrl manage the local system in relation with external devices (DMA) and operating system handshakes

adopting this common flow. The system behavior as it has been explained provides execution hypotheses that fix requirements for local processing specifications:

- Multi-processing: a group of processes is the way to allow concurrent accesses to memories
- Instruction level parallelism: allows to express concurrent group of operations inside processes
- Inter-process communications implemented as blocking directional channels
- Local memory access capabilities
- Control structures as found in high level programming languages
- Start and termination supports

Other constraints came from the necessity to allow cooperative programming and multiple forms of specification as inputs. In MORPHEUS, we had, as example, SPEAR methodology, C for computation specification, SpecEdit for state machines, etc. These constraints led to the conclusion that an abstract format was necessary, and this format would be acceptable for different partners, tasks, and development platforms.

A *Control Data Flow Graph* (CDFG) format was elaborated to support computation specification. CDFG has evolved during the first part of MORPHEUS project and is now stabilized and used as an interchange format allowing to assemble input process specifications, and to synthesize circuits for at least two very different targets: M2000 fine grain, and PiCoGA coarse grain.

CDFG structure has been influenced by languages from the *Communicating Sequential Processes* (CSP) family [1]. It is formalized into a data structure *model* holding nodes for atomic constructs, and hierarchy levels. The model is specified in the Express language coming from the STEP ISO project. Given this model, tools allow to automatically generate inter-change files formats and Application

Programming Interface (API) to write and read these files. Practically, free software Platypus [3] platform has been used to specify the model and generate APIs for a set of development environments including C/C++, Java and Smalltalk (Fig. 13.8).

To summarize CDFG contents, a compiler front-end can produce specifications for one process or a group of processes composed by channels. When writing a program graph to a CDFG, the compiler can set up:

- Memory read and write nodes
- Conditional, static and dynamic loops
- Arithmetic operation nodes
- Parallel constructs terminated by local barriers
- Multiple branch
- Hierarchy calls
- Channel based blocking communications

As it is, the CDFG development platform allows the composition of processes coming from different sources. A nice property of CDFG is that it can support simulation, before synthesis, allowing to collect execution characteristics not only at local level, but also in conjunction with system communication simulation support, allowing quick investigations on the accelerator data loop before making the choice of a particular target.

Obviously, code generation for a sequential processor is another opportunity, offering a bottom line for performances before further explorations.

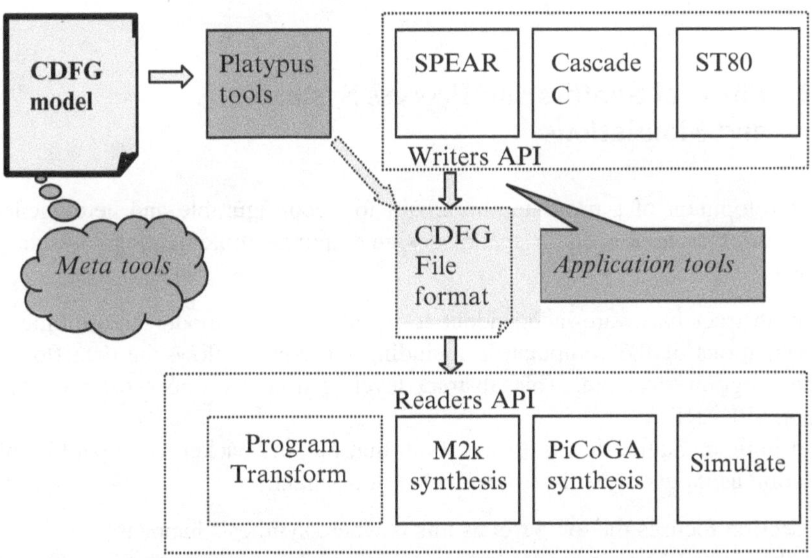

Fig. 13.8 CDFG abstract model, APIs and effective format exchanges

13.1.8 Discussion and Acknowledgements

The Spatial Design prototype received the contribution of J-C Le Lann with a system-C model for a multi-channel multi-context feeding engine operating on local buffers. The System-C code was assembled with two other parts:

- A C code representing the main program thread calling the accelerator. This code set-up a communication queue suitable for the communication engine, then activates this engine.
- A C code emulating sequentially the accelerated computation on the buffers filled by the communication engine.

This former experiment was complemented by an investigation on source to source transformations of C programs to separate automatically data access in main memory that goes to the communication engine command blocks, and the actual computation that goes to the accelerator (investigation from R. Keryell and M. Godet with the PIPS source to source compiler).

Feeding using DMA, overlapping communication and computation has yet been used for several accelerator boards. The multi-process implementation in a SoC and the muti-target, multi-source is a novelty. Another interesting point is the data feeding mechanism. In the case where enough buffers are locally available, it is possible to relax the strict synchronism between DMA queue execution and local computations. The local controller can evolve to allow data to flow in advance of the computation and can also delay result forwarding. Asynchronous communication model as used in variant of message passing interfaces can be reproduced, allowing the unification of MORPHEUS accelerators and parallel computing paradigms.

13.2 Physical Synthesis of Process Systems and Simulations

The development of a programming chain for reconfigurable and heterogeneous architectures as addressed in spatial design requires implementing tools in two levels:

1. An abstract hardware-independent level, allowing to model algorithmic representations of the computation including the control flow, the data flow and the program structure. This abstract level is used by upper software layers (e.g. SPEAR).
2. An implementation level, or low level, that outputs backend compatible information as inputs of hardware-dependent tool-chain.

This section focuses the low level as this is where synthesis happens.

In the Spatial Design scope, linking these two levels happens through sharing an application description formalism (CDFG intermediate representation) used as an information model and through its delivery within several programming environments. This provides an easy way to build up a tool suite, with neither black boxes nor semantic loss during application definition interchange.

The goal of an information model, as opposed to a data model, is to capture the concepts in a given domain of interest; it is a formal specification of that domain, that captures not only the objects, attributes and relationships in the given domain but also the constraints that apply to those objects and relationships. Such an approach clarifies many details early in the design process.

The lower level in the Spatial Design flow addresses tools supporting program transformations over the CDFGs with regards to the implementation targets specificities. In that way Madeo+ fills the gap between the partners' front-end tools and the MORPHEUS heterogeneous execution support.

In addition, simulation facilities are provided that offer a simple path to validation of the implemented applications.

13.2.1 CDFG Implementation HREs and Synthesis Principles

In the Spatial Design, two targets are under consideration: M2000 FlexEOS fine grained eFPGA, and DREAM medium grained reconfigurable platform from ARCES. These two HREs are very different in kind as the DREAM platform embeds a host processor to control the reconfigurable matrix of processing elements. Despite common hypothesis regarding the execution model apply whichever target is considered, these differences lead to develop two synthesis approaches.

Synthesis for eFPGA refers to netlist outputting. Netlists, that embed a controller for external synchronization, are further accepted by M2000 back-end tools. Synthesis for DREAM refers to code generation. This code is further accepted by DREAM compiler tool chain from ARCES.

13.2.2 CDFG Implementation in DREAM HRE

The DREAM platform is composed of two execution engines that are a 32-bit RISC processor and the PicoGA reconfigurable array [4,5]. This latter is connected to an interconnection matrix interfacing a set of address generators (AG), each coupled to a data exchange buffer (DEB). AG allows skipping, overlapping and power-of-2 modulo addressing.

The RISC processor is programmed by classical C code linked to the DREAM API for controlling the interconnection matrix configuration, AGs configurations, loading and triggering accelerated functions on the PicoGA. Accelerated functions are programmed in a subset of C, called Griffy-C, used as a format for describing behavioural data flow graphs [6].

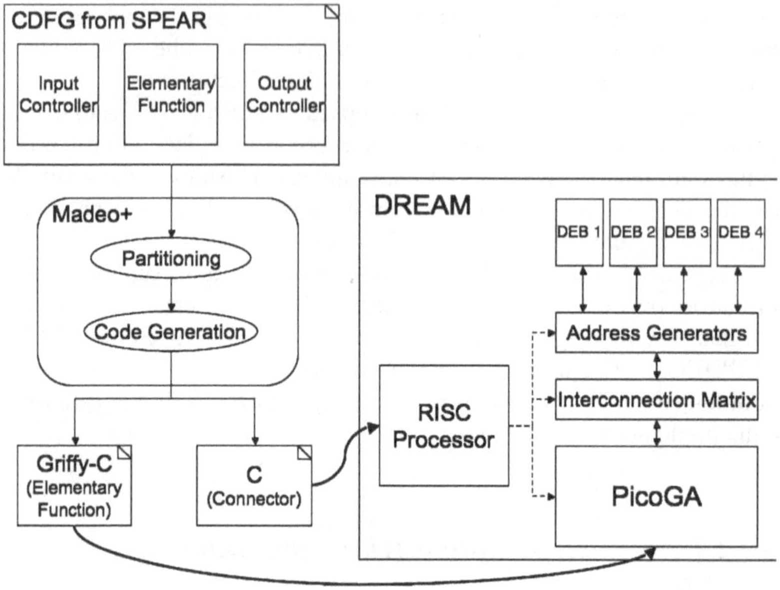

Fig. 13.9 CDFG implementation on the DREAM HRE

Madeo+ partitions a SPEAR CDFG in order to generate both codes for the RISC processor and the PicoGA (see Fig. 13.9). Application control parts are implemented on the RISC as nested loops. Their role is to compute configurations for AGs and to trigger accelerated functions. Griffy-C code is generated from the CDFG specification of the elementary function process.

Additionally to AG configuration and computation activation, RISC processor code initializes the interconnection matrix configuration (accelerated function's IOs as DEBs). The channel communications described in the high-level CDFG appear as DEBs read/write operations/AGs.

13.2.3 Execution Model and Scenario

The RISC is a mono-processor hence the parallelism expressed in the CDFG execution model is not supported. Figure 13.10 illustrates an example of CDFG with three processes executed in parallel and synchronized by communications.

Madeo+ applies transformations on the CDFG, such as *loop fusion* on the controllers, in order to match a sequential execution while the original semantic of each process is preserved. The C code generation follows a generic template compliant to the execution flow of the DREAM HRE described as follow:

1. System DMA writes input data in the DREAM's DEBs
2. The processor starts its execution with data located in the DEBs

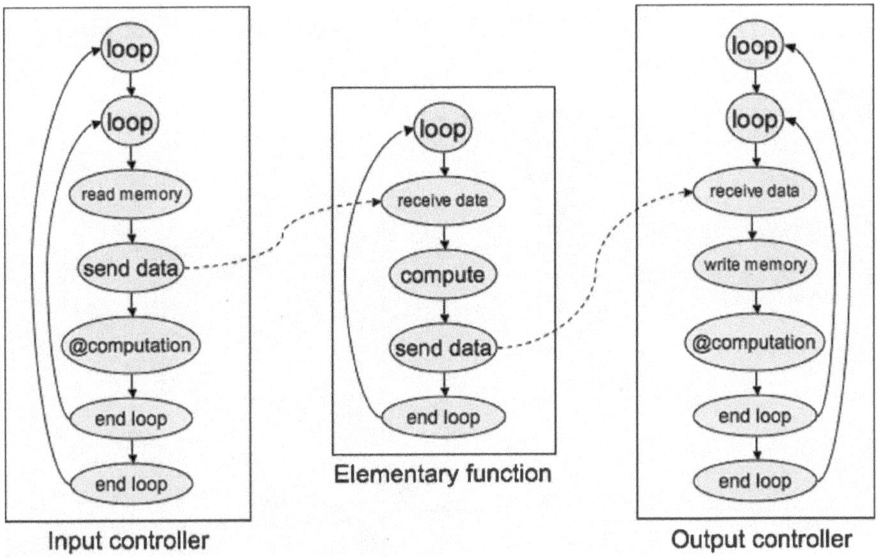

Fig. 13.10 CDFG execution model

3. Allocation of the accelerated functions on the PicoGA
4. Interconnection matrix configuration
5. Connector loop
   (a) AG configuration
   (b) Accelerated function activation and results written in a DEBs
   (c) Synchronization barrier
6. PicoGA deallocation
7. System DMA reads the results located in the DREAM's DEBs

The step "Connector loop" corresponds to the fusion of the input and output controllers of the CDFG since they cannot be implemented as parallel activities.

13.2.4 Accelerated Function Implementation

While the application control part is placed on the RISC processor, the elementary function of the application is placed on the PicoGA reconfigurable array as illustrated by Fig. 13.9. This computational part is implemented as a Griffy-C description generated from the CDFG. Griffy-C is a subset of the ANSI C syntax for describing Data Flow Graphs (DFG) to be implemented on the PicoGA. Restrictions mainly refer to control instructions, arithmetic operators and single assignment, that is variables are assigned only one time. Figure 13.11 illustrates an elementary function (a FIR) and its corresponding Griffy-C code.

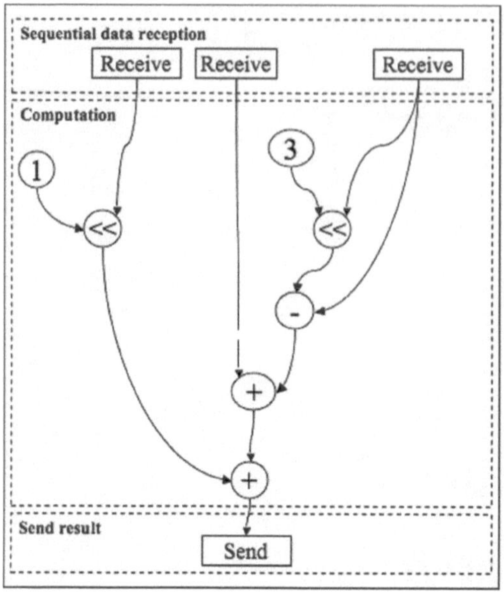


```
#pragma picoga processnode 1 1 output input
{
static int t1;
static int t2;
static int t3;
int temp_2236;
int temp_3845;
int temp_2534;
int temp_3247;
int temp_4149;

//Sequential data reception
t1 = t3;
t3 = t2;
t2 = input;

//Computation
temp_3845 = t3 << 1;
temp_2236 = t2 << 3;
temp_2534 = temp_2236 - t2;
temp_3247 = temp_2534 + t1;
temp_4149 = temp_3845 + temp_3247;

//Send result
output = temp_4149;
}
#pragma end
```

Fig. 13.11 CDFG elementary function and its corresponding Griffy-C generated code

13.2.5 Connector Implementation with the DREAM API: Principle

Two DREAM API functions are used for connector implementation:

- *Accelerated function activation*: *pga_op_x(idf, idm, iter)* with *idf* the identifier of an accelerated function, *iter* the number of activations and *idm* the interconnection matrix.
- *AG configuration*: *set_DF(item, addr, count, stride, step, rw)* with *item* the AG's identifier, *addr* the initial R/W address, *count* the number of data word per data chunk. *stride* the distance between data chunks. *step* the data accessing step and *rw* the R/W mode.

Figure 13.12a gives an example of C pseudo-code for an input controller similar to the one depicted by Fig. 13.11.

Figure 13.12b gives the generated RISC pseudo-code for the input controller. The address computation at loop-level 2 corresponds to the reading *step*. Similarly, address computation at loop level 1 is equivalent to the *stride* parameter. These parameters are both used for generating the AG configuration 2-D accessing pattern. The *starting address* is given by the initialization value of the address.

In order to produce a full connector both output and input controllers are generated as AG configurations (Fig. 13.13). If controllers have more than two loop-levels, extra loop-levels are not implemented by AG configurations but loop fusion is used to preserve increment on addresses.

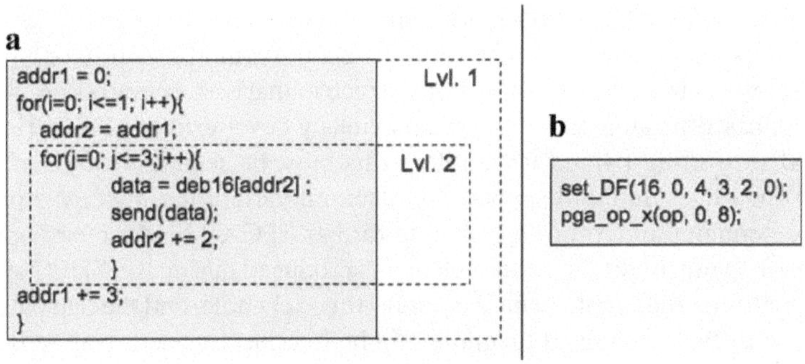

Fig. 13.12 (**a**) C pseudo-code of an input controller. (**b**) Equivalent AG configuration for the input controller

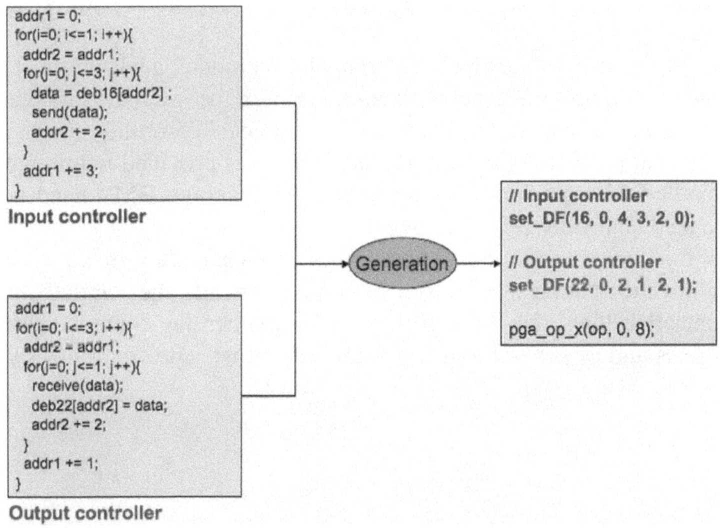

Fig. 13.13 AG configurations for two controllers and accelerated function triggering

13.2.6 CDFG Implementation in M2000 FlexEOS HRE

Not only spatial design aims at covering several reconfigurable architectures, but also the tools suite's ability to evolve in order to adapt to new architectures appears as a key issue. This claim for target evolution has been exercised within the scope of MORPHEUS project itself as M2000 offered two families (namely F4 and F5) of its eFPGA architecture.

The MADEO+ tool suite supports both primitive support (library of operators) and soft macro support (operator generators, allowing producing tailored operators based on parametric specifications).

In the first case, the CDFG operations are mapped using the operator library, with the closest type that covers the inputs. On the contrary, the F5 family only supports look-up tables and asynchronous flip-flops, hence in that case operators are generated on demand, based on an extension of the previously developed MADEO framework.

Beyond supporting F4 and F5 families, this must be seen as a proof of concept that MADEO+ does and will support architectural variability; hence the presented results are perennial in term of tailoring to further FPGA-like hardware options.

Figure 13.14 illustrates the various elements produced during the CDFG synthesis: the local controller that implements a compile-time schedule, and the datapath, which is composed of the accelerated functions and half-connectors acting as AGs.

The computation scheduling is made predictable by decoupling the functions from the DMAs, hence a timing analysis leads to a timing generator definition. The timing generator is in charge of dependencies resolution between memory accesses (generated from SPEAR) and the accelerated function itself (Task).

Synchronization channels are implemented as registers with a scheduled *enable* port.

In order to conform to the pipelined computation model, a barrier mechanism is implemented to support the synchronization between the internal computation and the DMAs. After one barrier has been crossed, the local controller in charge of the computation starts. As no specific hardware support is provided within the M2000 HRE to handle DMA synchronizations, MADEO+ generates DMA handshakes and a multiple-step controller (Fig. 13.15).

Each of the three activities rises up a termination signal. They are all consumed by the synchronization barrier mechanism. After n barrier calls, the computation is over.

The controller relies on a loop, in charge of both counting down the requests the DMA initiates and of generating the acknowledgments. The controller's synthesis

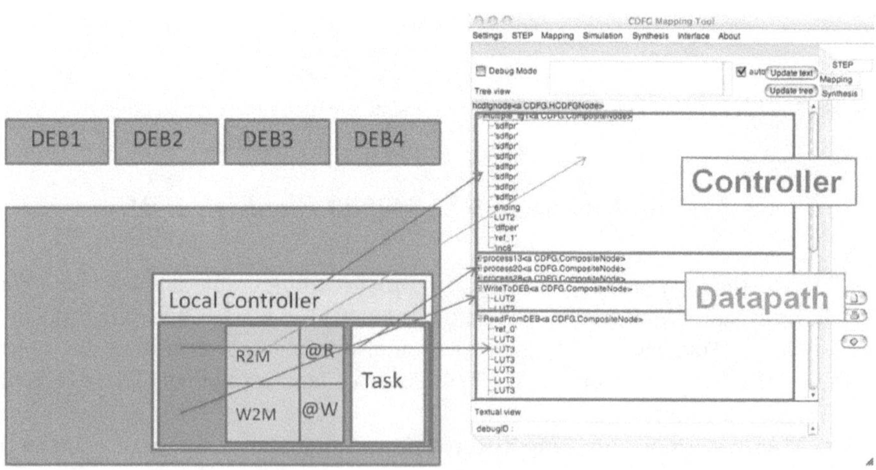

Fig. 13.14 First step, the CDFG is synthesized regardless to system activities

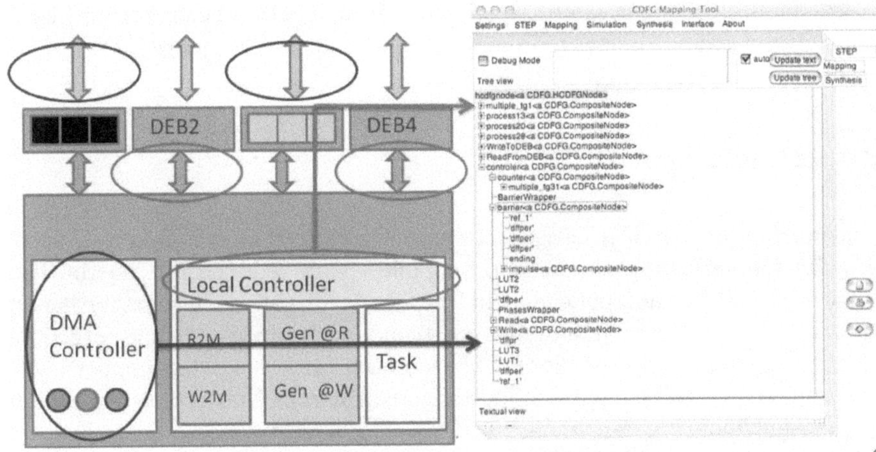

Fig. 13.15 The two levels of controller: local execution and DMA handshake

relies on logic synthesis and on simple and mixable circuit patterns (e.g. a loop and a multiple value threshold detector for local controller).

13.2.7 Tool Flow

The CDFG implementation starts with a High-Level which is further added IOs port, each of which links a DEB with a high level memory.

The synthesis stage generates a low-level CDFG, made up of several processes: the computation itself, the memory reader, the memory writer and the local controller.

In addition, the memory IOs accesses are integrated in the top-level interface, with respect to source-consumers. This point constitutes the memory harness for interfacing the DEBs ports.

At this point, the CDFG is decoupled from the three-stage pipeline model; the designer provides the number of steps to generate the controller.

This extended low level CDFG is converted in an EDIF RTL netlist, to be taken as input by the backend FlexEOS tool, in charge of generating reports (timing, allocation, etc.), Verilog code (useful for validation with MODELSIM) and possibly bitstream.

13.2.8 Simulation

Validation appears at two levels: the low-level CDFG is stressed to output execution traces allowing early validation before going through some mainstream (e.g. ModelSim) or dedicated tools (e.g. DREAM Simulator).

Assertions/breakpoints can be set, in order to run the simulation cycle by cycle until a given condition comes to true; what favours agile debugging.

13.3 Conclusions

Standardizing the use of heterogeneous reconfigurable resources in a SoC scope goes through providing a one stop shopping point (e.g. CDFG intermediate format) to describe the application independently of the implementation target, but with respect to programming (e.g. concurrent processes) and execution models (e.g. decoupled execution). This approach favours hiding to upper layers the specificities of both the execution supports and both their dedicated software environments/tools.

In the scope of the MORPHEUS Spatial design, Madeo+ delivers this facility and offers a path from a neutral application description format down to back-end compatible post synthesis flow.

References

1. C. A. R. Hoare ed., *Developments in Concurrency and Communications*, Addison-Wesley, Longman Publishing Co., Inc., Boston, MA, USA 1990.
2. S. Yazdani, Concurrent coordinated shared memory accers for reconfigurable multimedia accelerators. PhD thesis. Université de Brest, France 2008.
3. A. Plantec, Exploitation de la norme STEP pour la spécification et la mise en oeuvre de générateurs de code, Ph.D. Thesis (in French), Université de Rennes 1 and UBO, 1999.
4. F. Campi et al., A dynamically adaptive DSP for heterogeneous reconfigurable platforms, *Proceedings of the Design Automation and Test (DATE)*, 2007.
5. C. Mucci et al., Implementation of AES/Rijndael on a dynamically reconfigurable architecture, *Proceedings of DATE* 2007.
6. C. Mucci, Software tools for embedded reconfigurable processors, Ph.D. Thesis, University of Bologna, Italy, 2006.

Part IV
The Applications

Chapter 14
Real-Time Digital Film Processing

Mapping of a Film Grain Noise Reduction Algorithm to the MORPHEUS Platform

Henning Sahlbach, Wolfram Putzke-Röming, Sean Whitty, and Rolf Ernst

Abstract Real-time post processing for digital cinema is an extremely challenging task, due to large resolutions and the resulting high data rates. Applications with these requirements are beyond the scope of standard DSP processors, and ASICs are not economically viable due to a small market volume. As an answer to these challenges, the MORPHEUS platform offers reconfigurable processing engines with mixed granularity and an integrated toolset for rapid application development. This chapter presents a sophisticated film grain noise reduction algorithm and its mapping to the MORPHEUS platform.

Keywords Application • image processing • digital cinema • real-time • high performance • FlexFilm • noise reduction

14.1 Introduction

In recent years, motion picture studios and advertisement industries began demanding real-time or close to real-time processing to receive immediate feedback during interactive film processing. Post processing applications commonly required by these customers are characterized by resolutions of at least 2K ($2{,}048 \times 1{,}556$ pixels per image). The real-time requirement, together with complex image processing algorithms, results in high memory and computational needs that necessitate an accelerated implementation on dedicated hardware. Today, the image processing algorithms run offline on high performance standard PCs, which are significantly slower than the film scanners delivering the input images with a data rate of up to 15 Gbit/s.

H. Sahlbach (✉), S. Whitty, and R. Ernst
IDA, TU Braunschweig, Germany
sahlbach@ida.ing.tu-bs.de

W. Putzke-Röming
Deutsche Thomson OHG, Germany

N. Voros et al. (eds.), *Dynamic System Reconfiguration in Heterogeneous Platforms*,
Lecture Notes in Electrical Engineering 40,

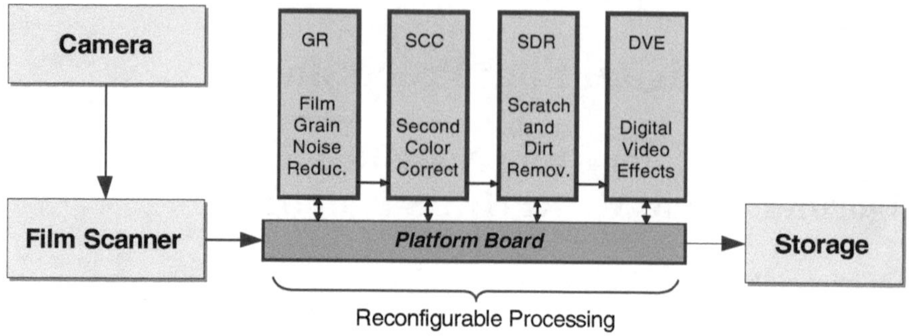

Fig. 14.1 Structure of a digital film post processing unit

Therefore, the preferred solution is (one or more) reconfigurable accelerators connected to the film scanner via a PC that perform the various processing steps.

Figure 14.1 shows the typical first phase of a post processing work flow, consisting of four different image processing algorithms:

- Film grain noise reduction
- Secondary color correction
- Scratch and dirt removal
- Video effects – rotation and spatial correction

The major goal of these algorithms is the normalization of the film material by compensating for effects introduced by cameras and film scanners. In context of MORPHEUS, film grain noise reduction has been selected from the four algorithms for hardware acceleration.

14.2 Film Grain Noise Reduction

Film grain noise is a photographic effect resulting from the granularity of film. As complete removal of this noise produces artificial looking images, a controlled reduction, or even the addition of noise for certain scenes are desired capabilities. Film grain noise reduction is a combination of different image processing algorithms that are representative of the entire video processing domain.

Figure 14.2 depicts the complete structure of the implemented algorithm [1], originally designed during the FlexFilm project [2]. The first step is a bidirectional motion estimation (ME) using a block-matching, exhaustive search algorithm that operates in the luminance space of the input images. The decision criteria used to select the optimal motion vectors is the minimization of the sum-of-absolute-differences (SAD). The resulting motion vectors are consumed by the motion compensation (MC), which builds an image out of image blocks from the preceding and succeeding images. This optimizes the noise reduction results, as it can better deal with object occlusion and image sequence cuts.

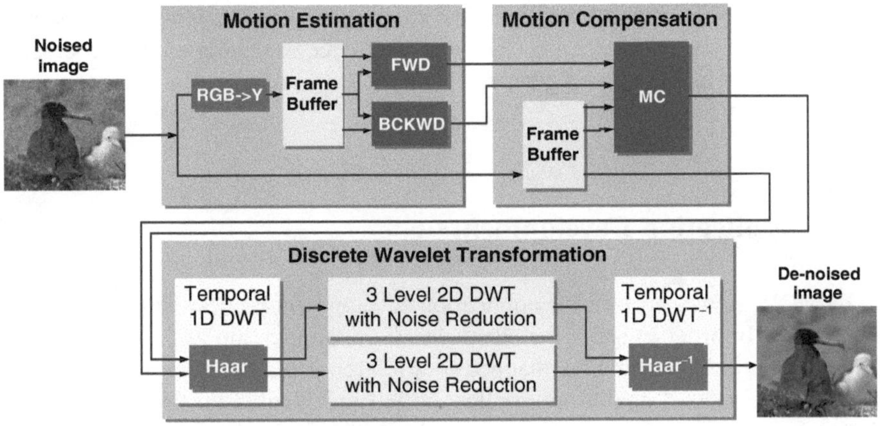

Fig. 14.2 Film grain noise reduction algorithm

The last part of the algorithm is the noise reduction itself, which is performed in wavelet space. The reference, together with the compensated image stream, is transformed in temporal dimension using a Haar wavelet, followed by a three-level discrete wavelet transformation (DWT) using a 5/3-wavelet [3], which operates in the horizontal and vertical directions. In wavelet space, noise reduction is performed via a shrinkage function with independent run-time configurable thresholds on all decomposed sub-bands. Afterwards, an inverse discrete wavelet transformation is applied to all sub-bands, reversing the 5/3-wavelet. The final inverse Haar filter recreates the de-noised versions of the original streams and completes the algorithm.

14.2.1 Application Characteristics

The film grain noise reduction application was selected for the MORPHEUS project for two reasons: first, the clear structure allows an easy partitioning of the various application parts to different processing engines. Second, each single part of the application has unique computational and memory requirements, which cover a large class of image processing algorithms.

In the ME unit, the computation of SADs is based on arithmetic operations and can be efficiently parallelized, leading to high computational density. The memory is accessed in a predictable fashion, offering the opportunity to perform stream processing. In contrast, the MC produces irregular block-based accesses on preceding and succeeding images, depending on the best match. Most operations are comparisons. Therefore, the MC does not require large computational resources. The DWT executes a mixed set of instructions ranging from multiplications, additions and shift-add operations. The bit width varies from 10 to 30 bits, requiring flexible data formats and memory accesses. Because of the multiple DWT levels and the various filter stages, the DWT is also computation intensive.

A detailed description of the algorithm can be found in [4]. However, this brief overview clearly demonstrates the very different requirements of the processing blocks, which makes the algorithm extremely suitable for a heterogeneous processing platform.

14.3 Application Development

In the MORPHEUS project, the application development process was organized in two phases. In order to show the adequacy of the application and to demonstrate its features, the algorithm was mapped to a reference platform in the first project phase. In the second phase, the application is implemented on the MORPHEUS platform, using valuable information obtained during the first mapping. Finally, the MORPHEUS implementation is compared to the reference implementation and the results are evaluated.

14.3.1 Mapping to the Reference Platform

The reference implementation is based on research performed in the successfully completed FlexFilm project [5]. The main goal of this project was the design of a new processing platform for high-performance image processing algorithms. During the project a component-based library named FlexWAFE [6] was designed, which introduces weak programmability for FPGA designs. Figure 14.3 shows

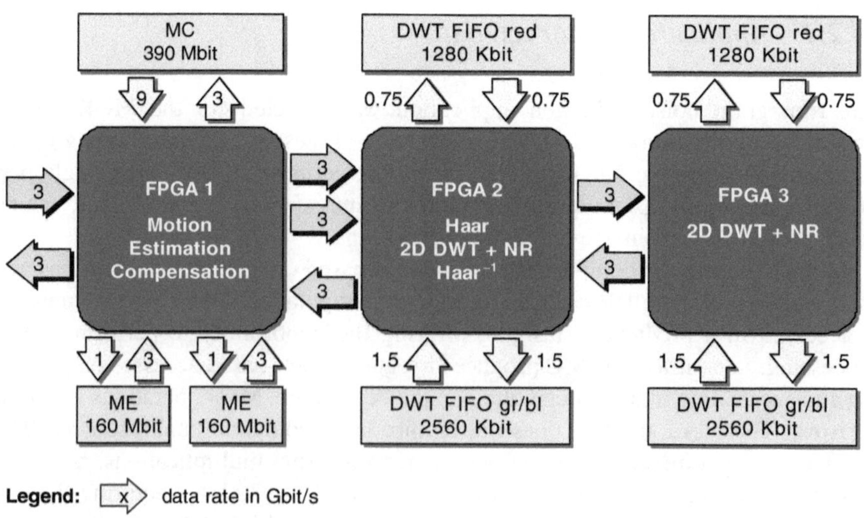

Fig. 14.3 Mapping to the reference platform

the decomposition of the application and its mapping to three Xilinx Virtex II Pro FPGAs (XC2VP50) of the FlexFilm board.

In the ME unit, which shares a single FPGA with the MC component, the calculation of the SADs was parallelized by constructing two systolic arrays, each consisting of 256 processing elements.

The remaining two FPGAs are consumed by the DWT component, which requires a large chip area due to increasing data widths of up to 30 bits per color component. For data transport, two different mechanisms were applied: inside the FPGAs, a stream-oriented three-signal protocol connects processing elements in the data path. The inter-chip connections use a TDMA-based protocol and multiplex several data streams into one channel. Both protocols support backpressure in case of pipeline stalls. High performance memory accesses are guaranteed by seven instances of a real-time capable memory controller [7,8].

14.3.1.1 Mapping Results

The key advantage of the reference platform are fast memory access capabilities provided by the custom designed memory controllers, which guarantee a data rate of 26Gbit/s and access 512MBytes per FPGA. Furthermore, each FPGA is equipped with 4.1Mbit of internal memory, which allows an on-chip buffering of complete image lines. This memory hierarchy allowed the design to avoid memory bottlenecks. As an FPGA consists of thousands of simple logic cells, it offers no predefined structures for data transport and control units. Therefore, custom solutions for each application part had to be created in order to satisfy the different data and memory requirements of the distinct modules. Although supported by the FlexWAFE library [6] and its sophisticated memory access patterns, this task led to a lengthy development cycle. Thus, the fine granularity is considered to be the main disadvantage of the FPGA approach.

With regard to performance, the originally proposed real-time requirement of 24 frames/s for 2K images was fulfilled and even exceeded by the reference implementation. The complete algorithm performs 2,000 operations per pixel and is therefore extremely computation intensive. The performance results are summarized in Table 14.1.

14.3.2 Mapping to the MORPHEUS Platform

During the mapping to the MORPHEUS platform, all major processing steps (ME, MC, DWT) have been ported to two processing engine simulators (PACT XPP, ST DREAM). This implementation approach was necessary to evaluate the characteristics of the heterogeneous processing engines and to obtain an optimal mapping for each application module. The final mapping decisions are discussed in the following sections.

Table 14.1 Performance of the reference platform

Resolution in Pixels	Frames/s	Data Rates	GOPS
1080p (1920 × 1080)	40	318 MBytes/s	167
2K (2048 × 1556)	26	318 MBytes/s	167
4K (4096 × 3112)	6.8	330 MBytes/s	173

14.3.2.1 Processing Engine Analysis and Application Mapping

Due to the heterogeneity of MORPHEUS platform, a detailed analysis of the each processing engine was conducted. The fine-grained M2000 eFPGA is very similar to the homogeneous FPGA board, since it also consists of a logic fabric supporting variable bit widths. However, it is the smallest processing engine and offers limited computation resources. Therefore, only a small application part of the noise reduction algorithm, such as the RGB2Y conversion is suitable for this unit.

The DREAM engine is designed as an array of multiple 4-bit cells that can be combined into larger data words. Furthermore, it offers fast reconfiguration capabilities of two clock cycles, which can be efficiently exploited by the DWT's multiple FIR filters executed consecutively in a row. In the MORPHEUS chip implementation, a limited version of the DREAM that only allows the execution of a single filter stage at a time is included. Therefore, run-time reconfiguration becomes essential for this processing engine. The DREAM is equipped with 2D DMA engines, which allow a frame-based memory access, but are not suitable for block-based accesses required by the ME and MC components.

Finally, the PACT XPP is based on a Kahn graph data streaming concept [9] and offers powerful 4D DMA engines, which support a sliding window memory access pattern. Such a pattern is useful for the ME and MC units to implement the required block-based memory accesses. Furthermore, a conversion of the image orientation from a row- to a column-wise pixel representation is supported, which is required by the ME implementation [10]. The ME can take advantage of the XPP's streaming concept, as it performs predictable memory accesses, allowing the composition of a gapless pixel stream. The fixed word width and a long reconfiguration time of approximately 1,000 clock cycles are the key disadvantages of the XPP, which turned out to be problematic for the DWT with its varying word width and multiple filter stages.

Based on this analysis and the experiences from the simulator implementations, the application is decomposed as shown in Fig. 14.4. The final mapping uses all available processing units. A more detailed presentation of the application's characteristics and its mapping to the MORPHEUS platform can be found in [11].

14.3.2.2 Memory Management

In the first project phase, the large and fast on- and off-chip memories of the FPGAs were a key factor to satisfy the application's exhaustive memory demands (2.3 Gbit/s per image stream for 2K images). As the implemented MORPHEUS chip is only equipped with a single memory controller and 256 Kbytes of on-chip

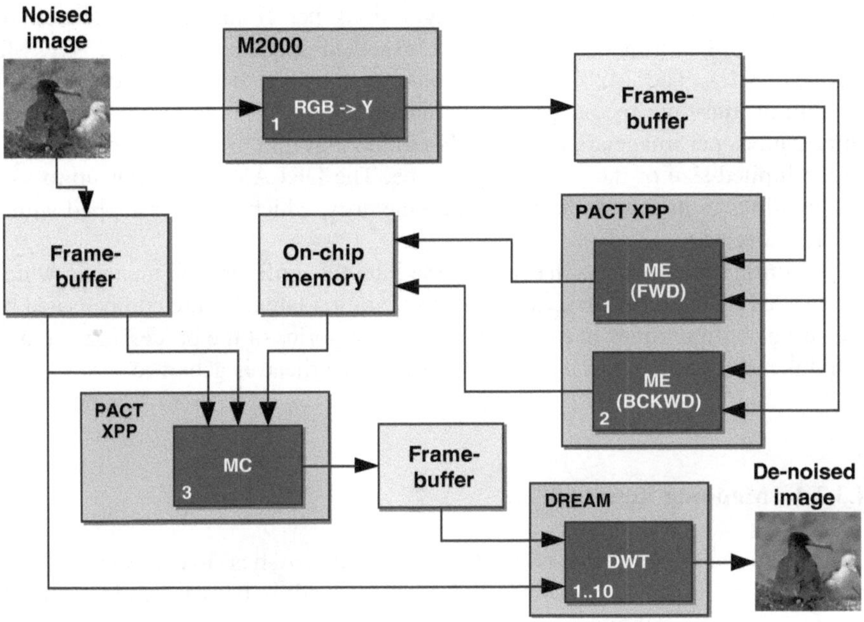

Fig. 14.4 Mapping to the MORPHEUS platform

memory, large resolutions like 2K images are out of the scope of the chip. In order to maintain an adequate frame rate for image processing, the image size has been reduced to SD-TV (720 × 576 pixels) resolution.

The amount of on-chip memory can be increased by converting the configuration manager memory into a data buffer. This memory can be directly addressed via the system's memory map and can be diverted from its intended use. Although this step reduces the amount of configurations that can be buffered on-chip, it might enhance the overall system performance. This trade-off between a larger amount of on-chip memory and faster reconfigurations needs to be quantified in upcoming experiments.

The memory architecture is completed by the processing engine's internal memories and caches, which are typically used for intermediate values or local variables. This heterogeneous memory hierarchy heavily differs from the FPGA's distributed, homogeneous on-chip memories and enforced completely new implementations of all application modules.

14.3.2.3 Run-Time Reconfiguration

As the FPGAs of the reference platform offered sufficient computation resources, a complete reconfiguration at run time was not necessary during the first project phase. For the MORPHEUS chip and its limited resources, run-time reconfiguration of some processing engines is required to support the application. In Fig. 14.4, the amount of necessary reconfigurations is annotated for each processing engine.

The PACT XPP needs three reconfigurations per frame as it is shared by the ME and MC components, which are executed sequentially (ME FWD, ME BCKWD, MC). The M2000 eFPGA only contains a single processing step, therefore no run-time reconfiguration is necessary. For the DREAM 10 different configurations per image are required, due to the multiple filter stages of the DWT and the limited size of the processing engine. The DREAM can buffer up to four configurations in its internal configuration memory, which can be swapped within two clock cycles.

The configurations are exchanged by the internal configuration manager, which is controlled by the ARM processor. Reconfiguration latencies are compensated by preloading configurations in the configuration memories of the processing engines. With this mechanism a minimal reconfiguration overhead is achieved.

14.3.2.4 Mapping Results

In general, the heterogeneous application characteristics match the different processing engines of the MORPHEUS platform and a complete and reasonable mapping of the algorithm is possible. Compared to the first phase, the development effort for the data transport mechanisms was reduced by specific MORPHEUS features, such as the XPP's 4D-DMA engines.

Unfortunately, the performance of the implementation was reduced by the small amount of on-chip memory and the single memory controller. That does not appear as a fundamental limitation as memory size scales well with technology and suitable memory interfaces are available for comparable technologies. For MORPHEUS, the parameters were reduced due to cost and complexity reasons.

The overall development time was significantly decreased (3 years vs. 1 year) in the second project phase. Performance numbers are not yet available as the chip is still under construction. The mapping results are summarized in Table 14.2.

Table 14.2 Mapping results for the MORPHEUS platform

	Motion Estimation	Motion Compensation	Discrete Wavelet Transformation
PACT XPP	• Stream-oriented • 4D-DMA • *Suited*	• 4D-DMA • *Suited*	• Slow reconf. • Fixed width • *Not suited*
ST DREAM	• Only 2D-DMA • Limited resources • *Not suited*	• Only 2D-DMA • *Not suited*	• Fast reconf. • Variable width • *Suited*
M2000 eFPGA	• Small size • RGB2Y possible • *Not suited*	• Small size • *Not suited*	• Small size • *Not suited*

14.4 Conclusions

This chapter presents the mapping of a film grain noise reduction application onto the MORPHEUS platform. Platform heterogeneity turned out to be suitable for the application. However, the current constraints required simplification of the complex algorithm. The general version of the MORPHEUS architecture has no major shortcomings and does not require such simplifications, making MORPHEUS a very reasonable platform for the application. The limitations exist primarily in the current version of the chip produced for evaluations purposes and may be overcome in future generations.

References

1. Eichner, S., Scheller, G., Wessely, U., Rückert, H. and Hedtke, R., 2005, Motion compensated spatial–temporal reduction of film grain noise in the wavelet domain, *SMPTE Technical Conference,* New York.
2. do Carmo Lucas, A., Heithecker, S., Rüffer, P., Ernst, R., Rückert, H., Wischermann, G., Gebel, K., Fach, R., Hunther, W., Eichner, S. and Scheller, G., 2006, A reconfigurable HW/SW platform for computation intensive high-resolution real-time digital film applications, *Proceedings of Design, Automation and Test in Europe (DATE)*, 194–199.
3. Le Gall, D. and Tabatabai, A., 1988, Sub-band coding of digital images using symmetric short kernel filters and arithmetic coding techniques, *International Conference on Acoustics, Speech, and Signal Processing* (ICASSP-88), 761–764.
4. Heithecker, S., do Carmo Lucas, A. and Ernst, R., 2007, A high-end real-time digital film processing reconfigurable platform, *EURASIP Journal on Embedded Systems, Special Issue on Dynamically Reconfigurable Architectures,* Volume 2007, pp Article ID 85318.
5. do Carmo Lucas, A., Heithecker, S. and Ernst, R., 2007, FlexWAFE – A high-end real-time stream processing library for FPGAs, *Proceedings of the 44th annual Design Automation Conference (DAC)*, 916–921.
6. do Carmo Lucas, A., Sahlbach, H., Whitty, S., Heithecker, S. and Ernst, R., 2009, Application development with the FlexWAFE realtime stream processing architecture for FPGAs, *ACM Transactions on Embedded Computing Systems, Special Issue on Configuring Algorithms, Processes and Architecture (CAPA).*
7. Heithecker, S., do Carmo Lucas, A. and Ernst, R., 2003, A mixed QoS SDRAM controller for FPGA-based high-end image processing, *Workshop on Signal Processing Systems Design and Implementation*, TP.11.
8. Whitty, S. and Ernst, R., 2008, A bandwidth optimized SDRAM controller for the MORPHEUS reconfigurable architecture, *Proceedings of the IEEE Parallel and Distributed Processing Symposium (IPDPS).*
9. Kahn, G., 1974, The semantics of a simple language for parallel programming, *Proceedings of the IFIP Congress 74,* North-Holland Publishing Co.
10. Sanz, C., Garrido, M. J. and Meneses, J. M., 1996, VLSI architecture for motion estimation using the block-matching algorithm, *EDTC*, 310.
11. Whitty, S., Sahlbach, H., Putzke-Röming, W. and Ernst R., 2009, Mapping of a film grain removal algorithm to a heterogeneous reconfigurable architecture, *Proceedings of Design, Automation and Test in Europe (DATE).*

Chapter 15
Ethernet Based In-Service Reconfiguration of SoCs in Telecommunication Networks

Erik Markert, Sebastian Goller, Uwe Pross, Axel Schneider, Joachim Knäblein, and Ulrich Heinkel

Abstract The development of next-generation telecommunication infrastructure meets a number of demanding challenges. Complex functionality, highest bandwidth demands, short development cycles and dynamic market requirements are combined with emerging technologies where standardization is not complete or subject to change. This involves a high risk of errors and non-conformances. Once the equipment is deployed, the update of chips is expensive and time-consuming, requiring re-spins of the devices and field exchange of circuit packs. Reconfigurable Systems-on-Chip will facilitate to accommodate changes subsequent to the deployment and thus significantly reduce modification costs. In this article we present a new reconfiguration technology for Systems-on-Chip in a telecommunication network: design parts of deployed devices are updated based on information distributed to the Systems-on-Chip with in-band Ethernet packet streams while the nctwork equipment stays in service.

Keywords Network node · reconfiguration · Ethernet

15.1 Introduction

The increasing complexity of communications systems, coupled with evolving standards and specifications pose tremendous pressure on the design process. Moreover, the telecommunication market is dynamic, requiring short development cycles. Manufacturers, which aim an early presence on the market, are not able to wait for standards and specifications getting stable and mature. As a consequence of standard and specification weaknesses, they have to cope with costly design re-spins as well as high project and business risks.

E. Markert (✉), S. Goller, U. Pross, and U. Heinkel
Chemnitz University of Technology, Germany
erik.markert@etit.tu-chemnitz.de

A. Schneider and J. Knäblein
Alcatel-Lucent, Nuremberg, Germany

N. Voros et al. (eds.), *Dynamic System Reconfiguration in Heterogeneous Platforms*, Lecture Notes in Electrical Engineering 40,

Fundamental constraints like performance, design costs, flexibility, power consumption, etc. usually result in a trade-off when determining the components for the implementation platform: ASICs, FPGAs, microcontrollers, μPs, DSPs, or a mixture thereof. Reconfigurable Systems-on-Chip (SoC) architectures offer a trade-off between the performance of ASICs and the flexibility of programmable general purpose processors. The key advantages are as follows:

- Higher performance and lower power consumption than pure FPGA solutions.
- Increased flexibility during development: late design changes and bug fixes may be incorporated even late in the development cycle.
- Shorter development cycle: boards including hardware and software may be developed and even manufactured before chip verification is finalized.
- Extended product life-cycles: manufactured devices can be adapted to changes of standards or customer specifications which were not foreseen during the design phase.
- Early market presence: time-to-market is shorter as there is no need to wait with development till standards and specifications are stable.
- Decreased design and maintenance cost: the number of time consuming and expensive design re-spins is significantly reduced.

To leverage these advantages, a methodology is required to extend typical signal processing ASICs by a small reconfigurable part. This enables the update of these SoCs in large networks without direct intervention of on-site service personnel. In MORPHEUS, the basic mechanisms are demonstrated to use the existing telecommunication infrastructure and standard network protocols for the distribution of reconfiguration data in a network.

15.2 Theory of the Approach

Once it is required to update SoCs deployed in a telecommunication network, there are basically three ways to transfer the reconfiguration data to the systems:

- *On-site download by service personnel*: each system is updated manually. This is not a real option as it causes an enormous effort to travel to all network sites.
- *Combination with a software release upgrade*: reconfiguration data are downloaded to the systems as part of a new software image. This requires internal communication channels from the main controller of a system to the various SoCs in that system. In currently deployed systems usually such channels are not in place. Further it restricts the opportunities to update SoCs to the long software release upgrade cycles. Many network providers, especially the large ones, deploy a new software release less than once a year.
- *In-band download within the communication signal*: unused bandwidth is filled with reconfiguration data and transmitted via existing data paths through the network. No additional internal interfaces are needed as the data stream goes through the SoCs anyway for signal processing. The reconfiguration data is

marked as such to enable the SoCs detecting and extracting their reconfiguration data from the data stream.

We consider the in-band download as the most appropriate method. It does not require any special hardware or software modifications, except the replacement of conventional chips by reconfigurable SoCs.

In the following firstly the update is explained from a network level view, whereas in the next section we describe what happens inside the chip. As communication protocol we chose Ethernet, but the method may be applied to any other kind of communication protocol like SDH/SONET, OTN, IP, Fiber Channel, etc. The selection of Ethernet for our demonstration was mainly driven by two factors: Ethernet is a convenient choice for proof-of-concept as it enables the usage of standard devices and is scalable down to a size reasonable for a demonstrator. Furthermore, Ethernet is one of the key protocols of future communication systems, rapidly evolving and thus providing enough weaknesses in standardization and specification of new features to justify the usage of reconfigurable SoCs.

For transmission with Ethernet, the reconfiguration data are split into several pieces, each one small enough to fit into the payload of an Ethernet packet. Figure 15.1 depicts the structure of such an Ethernet frame.

The reconfiguration data is located in the payload portion of the packet. It consists of several sections:

- The *Reconfiguration Data Header* marks the payload as such.
- The *Reconfiguration Device Address* selects the device to be reconfigured. A node can contain more than one reconfigurable device. These devices may need individual reconfiguration data due to different technologies (ASIC, FPGA) and/or different functions. So it is essential that every device in a single node can be addressed individually.
- The *Reconfiguration Packet Number* (Rec. Pkt #) is needed to be able to restore the correct order of the reconfiguration data portions. The reason for this requirement is the fact that the consecutive transmission of reconfiguration

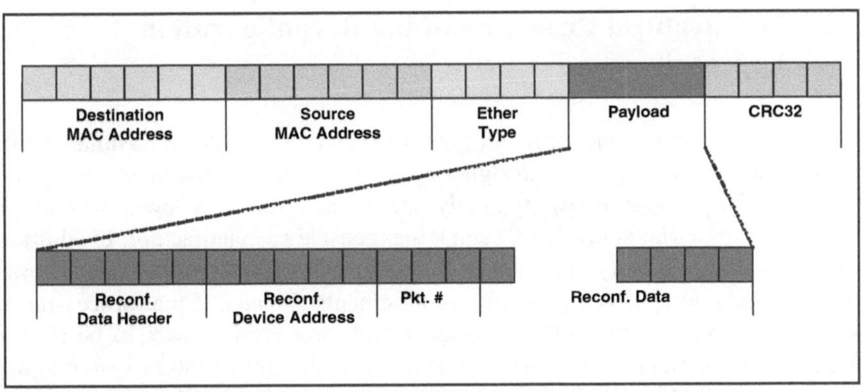

Fig. 15.1 Reconfiguration packet structure

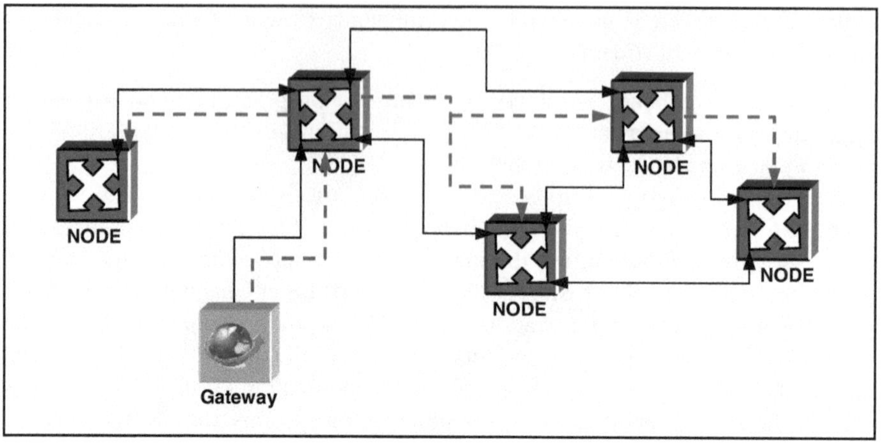

Fig. 15.2 Broadcast of reconfiguration data in a telecommunication network

packets through the network cannot be guaranteed. Therefore the data must be sorted before reconfiguration can start.

- The *Reconfiguration Data Payload* contains portions of the reconfiguration data, which are collected until the entire reconfiguration image is complete. After sorting and validation of the image the reconfiguration process is initiated.

The reconfiguration packets as described above are created at a central location and transferred to the network via a distribution gateway (for which e.g. the network management gateway may be used). As illustrated in Fig. 15.2, the gateway transmits the packets as broadcast to the whole network. Thus, in a properly configured network the reconfiguration data reach automatically all nodes with SoCs to be updated.

15.3 Architectural Principles of the Reconfiguration Approach

For communication systems different types of chip architectures are available with various characteristics, giving a designer a choice in selecting the most appropriate for a particular purpose. ASICs typically operate at a relatively low power and, if a large production run is involved, can be inexpensive to manufacture. In addition, they can be more densely provided in a system because of their relatively lower power consumption, reducing cooling requirements. Finally, signal processing in high-end communications systems requires high processing power to be able to handle single line data rates of up to 40 Gbit/s (and more in future systems) and total capacities in the multi Terabit/s range.

Thus, ASICs often are the preferred choice of a designer. On the other hand, FPGAs provide much more flexibility. The implementation may be corrected or updated anytime during design and verification – and even after deployment it is still possible to reprogram FPGAs.

The approach presented herein leverages the advantages of both variations by combining them into one System-on-Chip. For this purpose, the M2000 embedded FPGA technology is applied. The design parts, which are considered to be candidates for future changes ("weak parts"), are mapped to the embedded FPGA, whereas the design parts regarded as stable are implemented in the ASIC. These design decisions are essential and need to be a trade-off between keeping the embedded FPGA small and correctly identifying all weak parts. The architecture of such a reconfigurable SoC is illustrated in Fig. 15.3.

Once an Ethernet packet is received at the input of the SoC, it is first processed by a packet filter. The filter detects the reconfiguration packets addressed to this particular device and extracts it from the regular data stream. Reconfiguration packets sent to different types of devices are ignored and forwarded like normal customer data packets.

This way it is possible to have different kinds of reconfigurable devices in a network and to reconfigure them independently with separate data streams. The extracted reconfiguration packet then is duplicated. One copy is sent to the reconfiguration memory (RAM), where the reconfiguration data are collected. This RAM is additional memory dedicated for reconfiguration. Its size depends on the size of the reconfigurable core – which usually should be small compared to the ASIC part of the chip. The other copy of the packet is forwarded to the regular signal processing part of the SoC, as it is done for regular customer data. This broadcast ensures that the reconfiguration data are not absorbed by the first updated SoC but are continued to be forwarded to the rest of the network.

As soon as the reconfiguration data is complete – determined by packet number and checksum – the reconfiguration controller initiates and controls the update of the embedded FPGA.

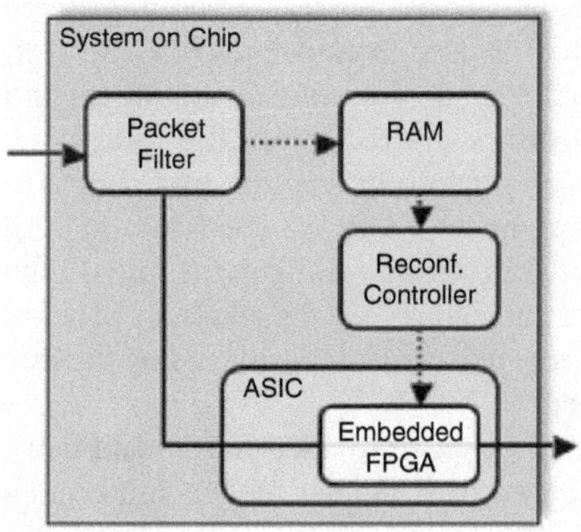

Fig. 15.3 Reconfigurable SoC architecture

15.4 Implementation

15.4.1 Platform and Tools

For demonstration purposes the node was emulated on a prototype platform consisting of two Xilinx boards (XUP Development Board). For the implementation of the prototype platform XILINX EDK 8.2i was used. The EDK provides a software development environment based on C language. Design specific libraries can be generated which allow hardware/software co-design. Furthermore the EDK allows the generation of scripts for behavioural, structural and timing verification of the design using Modelsim. Synthesis, place and route and the generation of configuration files are possible too. The EDK uses Xilinx XST for synthesis.

15.4.2 System Overview

Since the prototype platform emulates an embedded FPGA placed on an ASIC it has been divided in corresponding parts. The embedded FPGA macro contains an Ethernet MAC. The MAC receives an Ethernet data stream generated by an external PC. Each received Ethernet packet is checked for configuration data by searching a predefined source address in the header of the packages. If a reconfiguration packet has been detected the payload of the package is copied from the data stream and sent to a PowerPC core. To show the proper operation of the demonstrator it contains a loop functionality which sends back all received Ethernet packets to its source.

The MAC can be monitored and controlled by the PowerPC core. Software running on this core provides access to all registers of the MAC. A separate RS232 connection between the external PC and the board allows an Ethernet independent communication. So the functionality of the MAC can be checked even if the Ethernet connection has a malfunction.

After the configuration data has been received it is stored in the onboard DDR RAM. The PowerPC core is used to calculate the CRC of the entire configuration stream. If no error is detected, the reconfiguration of the embedded FPGA is initialized. Figure 15.4 shows an overview of the system.

The success of the reconfiguration is shown by fixing an initially faulty Ethernet MAC. Ethernet packets contain a CRC32 checksum to verify their payload. This CRC32 calculation produces wrong results in the initial Ethernet MAC.

To make this error visible, all received packets are sent back to their source. The CRC32 calculation error is detected and displayed by the PC. After reconfiguration the CRC32 error is fixed.

The embedded FPGA has been divided into a data path and a control path as shown in Fig. 15.5. The data path handles the incoming and outgoing Ethernet data stream and the identified configuration data. The MAC stores received and valid packets in a FIFO. If at least one packet is available the loop function starts reading

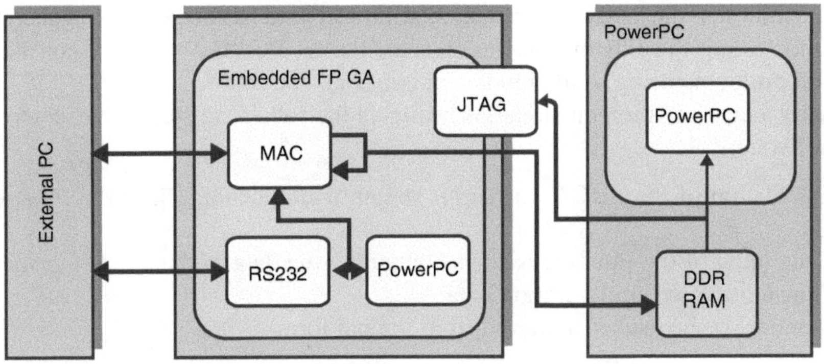

Fig. 15.4 Overview of the demonstrator platform

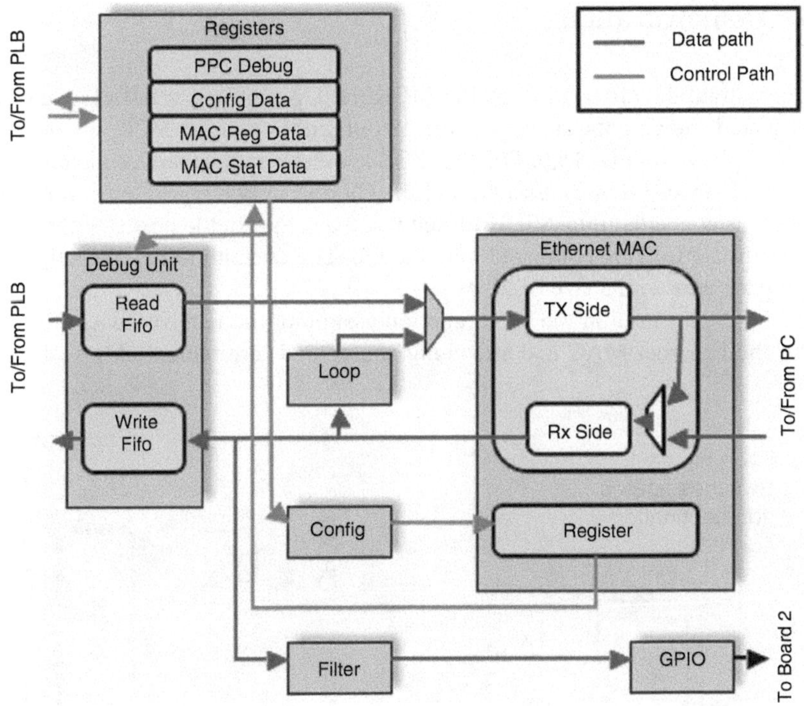

Fig. 15.5 Ethernet MAC wrapper

the data from the MAC. The source and the destination address are exchanged and the packet is assigned to the transmit path of the MAC (LOOP function). This path has to calculate the CRC32 for the outgoing packets before sending them to the PC. The control path provides access to the internal registers of the MAC that store all necessary parameters and statistics.

In addition to the hardware implementation software has been written generating and monitoring the Ethernet traffic on the external PC. The software consists of several programs being written in C and two shell scripts.

After a configuration bit stream is available the following steps are executed by the software:

1. Calculation of the CRC32 of the bit stream and appending it to the bit stream file.
2. Generation of the Ethernet packets containing the data of the bit stream file.
3. Transmitting the configuration data.
4. Monitoring the packets received from the platform.

The user can control the entire environment using the MORPHEUS-GUI.

15.5 Demonstration

To demonstrate the application on the MORPHEUS platform a test case has been implemented showing the success of the reconfiguration of the M2K macro during runtime as shown in Fig. 15.6. The FlexEOS synthesis tool generates the bit stream for the M2K macro (binary format) and the CRC32 in two separate files. First the bit stream is converted into ASCII format using a script. In the next step the stream data and the CRC32 are combined into one file. The content of this file is then split up into packages with a size of 20 bytes.

For the demonstration the loop function assigning the received data to the TX path of the Ethernet MAC and switching source and destination address contains

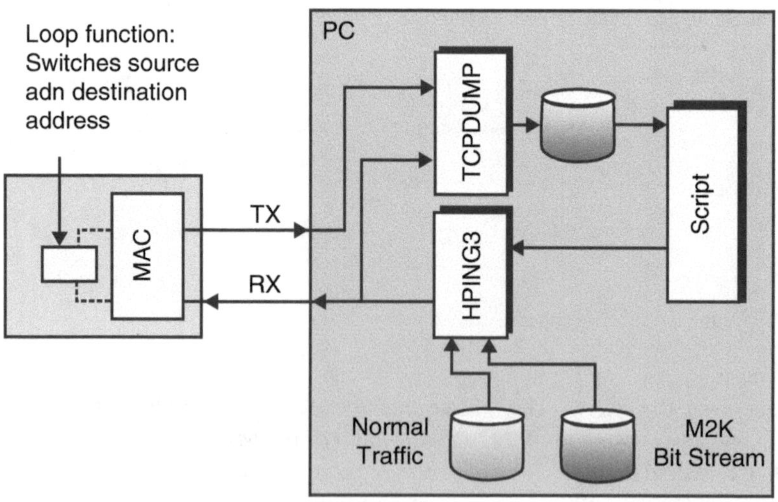

Fig. 15.6 Testmode for reconfiguration

an error in its initial version. The error produces a wrong source address during the switching procedure. The Ethernet traffic is generated using the program HPING3. The packages being sent from the external PC and received from the M2K macro are monitored with the program TCPDUMP. The output of this software is stored in a file. A script compares the destination address of the outgoing packages with the source address of the received packages. If they do not match the reconfiguration of the M2K macro is initialized. HPING3 uses the content of the bit stream file generating payload for the Ethernet packages.

The packages are received by the M2K macro and stored in DEB4. From there they are passed to the DREAM macro which calculates the CRC32 checking whether the bit stream has been received correctly or not. After the CRC check the data is stored in the configuration memory of the MORPHEUS device. The ARM processor is used to manage all processes on the MORPHEUS chip. It first waits until the M2K signalizes all data has been received. After that it checks the DREAM for finishing the CRC calculation. If the DREAM does not show an error the ARM starts the reconfiguration of the M2K.

During the reconfiguration process the error on the loop function of the M2K application is fixed. When finished the PC script comparing the addresses does not show any more errors.

Chapter 16
Homeland Security – Image Processing for Intelligent Cameras

Cyrille Batariere

Abstract Image surveillance systems demand is a challenging fast growing domain for TOSA [Thales Optronique Société Anonyme]. To address this domain and image processing architecture in general, TOSA bets on the strategy of using reconfigurable multi-purpose architecture to improve performances, re-use, productivity and reactivity. MORPHEUS project allows to realize this concept and demonstrate its capacities. Among the two phases that compose the MORPHEUS project the first phase is based on the implementation of a motion detection algorithm on an intermediate platform built by TOSA while the second phase is based on the implementation of the same motion detection algorithm on the MORPHEUS chip simulator. Some metrics are defined so as to compare the implementation on the two platforms with a traditional implementation scheme. A global work plan of how to implement any algorithm on MORPHEUS kind of platform is set up. After definition of this work plan, the implementation on phase one demonstrates the advantages and improvements of using this type of architecture while measures on phase two are not yet complete but should confirm the realization of the promises given by the platform.

Keywords Surveillance system · motion detection · image processing · SIMD · PiCoGA

16.1 Introduction

The domain targeted by TOSA [Thales Optronique Société Anonyme] for MORPHEUS is the emerging domain of intelligent optronic surveillance systems. Those systems have various functionalities and require several algorithms to be run to achieve the task. For instance such a system could be composed of a low

C. Batariere (✉)
Thales Optronics S.A., France
cyrille.batariere@fr.thalesgroup.com

N. Voros et al. (eds.), *Dynamic System Reconfiguration in Heterogeneous Platforms*,
Lecture Notes in Electrical Engineering 40,

consumption sleeping detection algorithm and an advanced set of recognition and identification algorithms in case of alarm. Achievement of such a system requires a flexible architecture to address a wide variety of processing, to allow fast and easy modifications of the application and to ease implementation of new functionalities. Classical implementation of hardwiring a lot of algorithms in VHDL onto FPGAs cannot be considered as an economically credible solution to implement such systems. For this reason, reconfigurable architectures such as MORPHEUS will be a breakthrough opening the way to new markets.

TOSA's contribution to MORPHEUS is divided in two phases. Both phases consist of implementation of a test case application on a platform and measurement of comparative performance figures. First phase consists of the implementation of a chosen application on an intermediate platform built by TOSA. Second phase consists of the implementation of the same application on the MORPHEUS chip simulator.

The first section of this chapter presents the context, the application and the MORPHEUS project. Second and third sections deal with the presentation and implementation of the application on each of the two platforms used during MORPHEUS project.

16.2 Application Presentation

16.2.1 TOSA Requirements

16.2.1.1 Industrial Context

The system targeted by TOSA is a general-purpose multi-applications image processing system. The motivation to move from application-dedicated architectures to general-purpose reconfigurable image processing systems is the following:
Traditional implementation of an image-processing algorithm consists in using a FPGA and hardwiring the algorithm in VHDL onto this FPGA. Such an approach is acceptable as far as only very simple image processing algorithms are considered. The main limitation of this approach is that there is more and more a strong customer demand for intelligent surveillance systems that are able to extract the relevant information from an image and to decide whether or not an alert should be raised. Such systems can be viewed as a large collection of real time algorithms, which are activated or not in function of non-predictable events such as the content of the image. For such systems, the above-mentioned traditional implementation of image processing algorithms is simply not feasible because too many different algorithms are required to implement an intelligent surveillance system.

In addition to all these considerations, for economical reasons a design goal should be to avoid designing a new application dedicated architecture every time a new application is required. Instead of that, the approach should be to design a single multi-application reconfigurable platform architecture on the top of which a lot of different image processing applications can be implemented.

16.2.1.2 Implementation Decomposition Through Three Levels

Algorithms are implemented as combinations of low-level building blocks (which will be called "operators" in the reminder of this chapter). A typical way to structure our application is to divide it into three layers as we can see in Fig. 16.1.

Such an implementation is convenient for a large class of image processing algorithms, namely those algorithms that can be viewed as sequentially applying a set of operators to an image.

Such an implementation demonstrates very strong reconfiguration capabilities:

- Capability to reconfigure into several different operators during the processing of an image so as to enable the implementation of complex image processing algorithms
- Capability to reconfigure into several different image processing algorithms so as to enable the activation of a given algorithm in function of a non predictable event such as the content of the image (intelligent behavior)
- Capability to use the same device for a lot of different applications and a lot of different customers

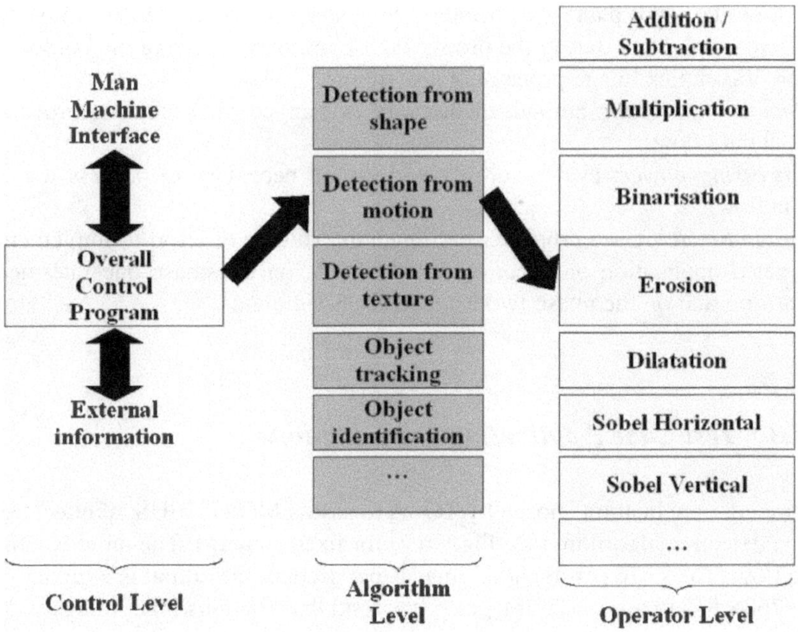

Fig. 16.1 Hierarchy in functions

In a typical implementation:

- Operators are run by the HREs on MORPHEUS and by a SIMD on a FPGA on intermediate platform.
- Algorithms are run by the ARM microprocessor on MORPHEUS and by a μBlaze on intermediate platform.
- Overall control is run by the ARM microprocessor on MORPHEUS and by a Pentium on intermediate platform.

Our modular approach gives the platform high flexibility, and enables easy reuse of existing functions for new applications. Building blocks are available both for high-level and low-level functionality.

16.2.2 Implementation Metrics

A set of metrics has been defined to evaluate the gains of performance and productivity by comparing implementation between the two platforms concerned by MORPHEUS project for TOSA (intermediate TOSA platform and MORPHEUS simulator) with traditional implementation when possible. This set of metrics is composed of:

- *In-process real-time reconfiguration*: number of reconfiguration per second to process the input data flow. It means the capability to reconfigure into several different operators during the processing of an image allowing the implementation of complex image processing algorithms.
- *Memory bandwidth*: amount of memory exchanged per second to process the input data flow.
- *Processing power*: cycles-number per second necessary to process the input data flow.
- *Design time/cost*: we propose to estimate the volume of effort to implement the targeted application on a conventional FPGA, on the phase one intermediate platform and on the phase two MORPHEUS-simulator.

16.2.3 Test Case Application Introduction

The test case application chosen by TOSA to assess MORPHEUS architecture is a motion detection algorithm (see Fig. 16.2) for fixed cameras. The input is a digital video flow (768×576 pixels) at 25 images per second, the output is a digital video flow (768×576 pixels) at 25 images per second that contains input image with the boundaries of moving objects underlined. The algorithm fits the intrusion detection in homeland security domain. The algorithm is divided into ten low level basic image processing operators.

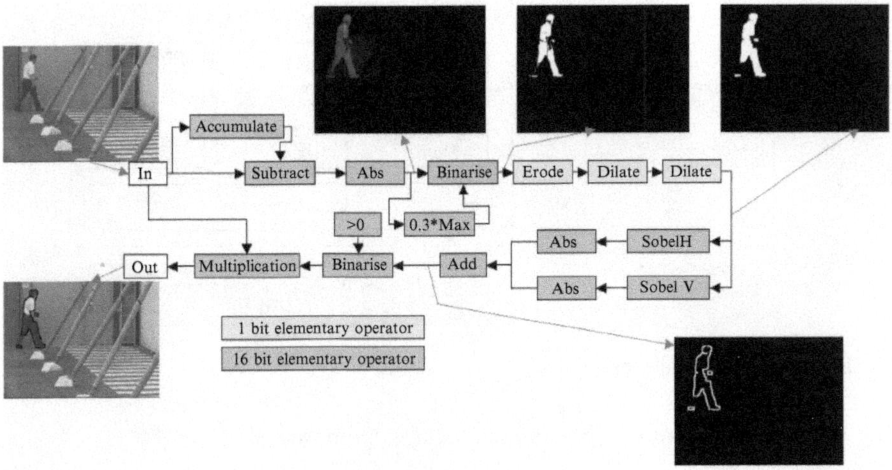

Fig. 16.2 Phase 1 motion detection algorithm

16.3 Phase One Achievements

16.3.1 Overview

Phase one study has been useful to analyze different ways to implement the motion detection algorithm and to get some figures about the metrics presented in the previous section on an intermediate platform used by TOSA. The preferred solution among these different ways depends on memory available for operators, memory available for data, memory bandwidth available. The different approaches are described in detail below.

16.3.2 The Intermediate Platform

The intermediate platform allowed TOSA to implement a demonstrator for an intelligent camera system that illustrates the usefulness of a reconfigurable system. This platform is very similar to the final MORPHEUS architecture.

From a hardware point of view, this intermediate platform is composed of a camera connected to a PC itself connected to a FPGA COTS board through a PCI-X interconnection.

The FPGA COTS board contains essentially (see Fig. 16.3):

- A Virtex4 SX-55 FPGA from Xilinx which is in charge of the processing
- A Virtex2 Pro FPGA from Xilinx which is in charge of the interface with the PC
- A DDR external RAM which provides the Virtex4 with a large storage capability

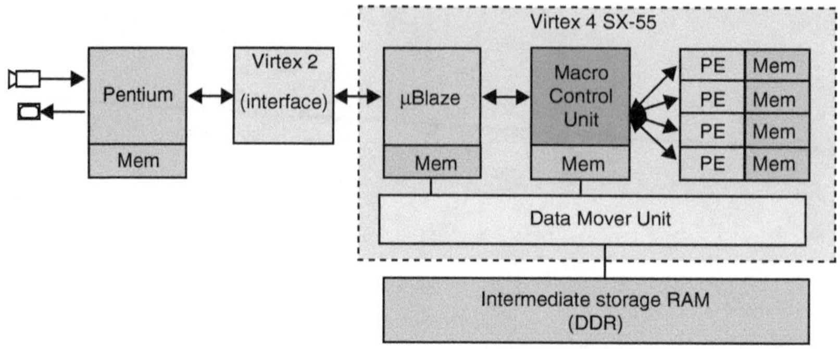

Fig. 16.3 Hardware architecture

The Virtex4 FPGA implements an architecture composed of:

- A μBlaze microprocessor in charge of the overall control of the chip and of activation of operators
- A SIMD containing a Macro Control Unit (MCU) and 128 Processing Elements (PEs) in charge of the bulk processing
- A Data Mover Unit (DMU) in charge of the transfers of data (principally images) between the storage RAM on one side and the PC, the μBlaze and the SIMD on the other side

16.3.3 Low Level Image Processing Operators

A basic set of image processing operators has been defined. This set is composed of low-level operators frequently used in image processing.

The set of operators used in phase one and partially in phase two is composed of (Table 16.1).

16.3.4 Implementation Philosophy/Schemes

In order to cope with hardware constraints (both in COTS and in MORPHEUS implementation), two main design drivers have been adopted:

- Image partitioning in tiles (full size image doesn't fit in processing memory).

Design considerations

- Size of tiles: square of 16,384 pixels tiles (128 × 128).
- Algorithm partitioning in operators clusters. An operator cluster is an agglomeration of one to several elementary operators.

Design considerations

- Each operator cluster must fit in the chip in terms of available gates number.
- Data memory amount for each operator cluster must fit into data processing memory.

Table 16.1 Set of operators implemented

Combination of two images
• Inf of two images (keeping the minimum of two images pixel by pixel)
• Sup of two images (keeping the maximum of two images pixel by pixel)
• Weighted addition (ax + by) of two images
• Multiplication of two images
Image2 = f (image1, coefficients)
• Upper threshold of an image
• Lower threshold of an image
• Linear transformation of luminance (ax + b)
• Division by 2^n
• Binarisation against a threshold
Filtering
• Convolution with a horizontal segment
• Convolution with a vertical segment
• Convolution with a rectangular window
Morphology
• Erosion (keeping the minimum value in the 3 × 3 neighborhood)
• Dilatation (keeping the maximum value in the neighborhood)
Statistics
• Minimum value of an image
• Maximum value of an image
Miscellaneous
• Copy of an image
• Absolute value of an image

Algorithm partitioning is an important point of optimisation work and this step is necessary for each new algorithm. Keeping these two design drivers in mind, different possibilities have to be considered to find the better scheme to implement any algorithm.

This paragraph's aim is to discuss the best implementation scheme for any "content dependent" algorithm. By "content dependent" algorithm, we mean an algorithm in which some operators are activated or not depending on information extracted from the image (typically results of previous operators).

Two different implementation patterns have been considered. The first implementation pattern consists in applying an operator cluster on each tile before applying the next operator cluster (left column of Table 16.2), the second implementation pattern consists in applying all the operator clusters on a tile before processing the next tiles (right column of Table 16.2).

The second implementation pattern has to be discarded for the following reasons:

- Reconfiguration rate is very high as reconfigurations occur for each tile.
- This implementation doesn't allow data dependent processing. For instance in our algorithm we need maximum value of the image to compute the first binarisation. As we need the maximum value of the whole image and not of a tile we can't apply the whole algorithm on a tile before processing the next tile.

Table 16.2 Two implementation patterns

First Implementation Pattern	Second Implementation Pattern
For all Operator_Cluster i	For all Pixel_Tile j
Load Operator_Cluster (i)	Load Pixel_Tile (j)
For all Pixel_Tile j	For all Operator_Cluster i
Load Pixel_Tile (j)	Load Operator_Cluster (i)
Compute()	Compute()
Next j	Next j
Next i	Next i

For the first implementation pattern various sub-pattern can be considered:

- Operator clusters are composed of just one elementary operator (which means 15 operator clusters). This is the first extreme case of our chosen implementation pattern.
- One operator cluster is composed of the whole algorithm (which means 1 operator cluster). This is the second extreme case of our chosen implementation pattern.
- The third case is a trade-off between the two above mentioned cases (which means 2 to 14 operator clusters).

The first sub-pattern has been discarded for the following reasons:

- This case consumes a lot of memory bandwidth as the tiles are transferred for each elementary operator (856,000 Kbytes/s).
- Reconfiguration rate is medium (375 reconfigurations/s).

The second sub-pattern has been discarded for the following reasons:

- This sub-pattern doesn't allow data dependent processing.

Finally the third sub-pattern of the first implementation pattern has been selected. Considering our algorithm, a natural choice is to have three operators clusters ($N = 3$):

(a) Accumulation, Subtraction, Absolute Value and Maximum Extraction
(b) Binarisation, Erosion, Dilatation, Dilatation, Sobel horizontal and vertical (each sobel operator stores absolute value of the result)
(c) Addition, Binarisation and Multiplication

Such a sub-pattern allows to conciliate:

- Moderate reconfiguration rate (75 reconfigurations/s if algorithm is composed of 3 operator clusters).
- Moderate memory bandwidth (276,000 Kbytes/s if algorithm is composed of 3 operator clusters).
- Possibility to tune to hardware. This sub-pattern allows adapting the implementation to the processing unit performances figures. If memory bandwidth is low and reconfiguration time is low or medium we can choose to minimize bandwidth by applying operator clusters composed of a maximum of elementary operators on a tile before computing the next tile. If bandwidth is high and

reconfiguration time is high we can choose to minimize reconfigurations by maximizing number of elementary operators in each operator cluster.

In addition this sub-pattern allows high gain in re-usability and productivity with the possibility to define operator clusters in terms of processing functions. For instance we can imagine a "segmentation" cluster or a "detection" cluster that is fully compatible with the three execution levels presented in Section 16.2.1.2.

16.3.5 Implementation Results Against Metrics

- Reconfiguration rate depends on number of operator clusters and frequency. For each image we have 3 reconfigurations and we have 25 images per second, the reconfiguration rate is 75 reconfigurations per second.
- Processing power.

Let's introduce the metrics used in Table 16.3:

- *Number of operator calls*: Number of operator calls to achieve the whole motion detection algorithm.
- *Cycles number per image processing*: This is the number of cycles needed by the SIMD to process the whole tiled image multiplied by the number of operator calls.
- *Cycles for an image per pixel*: This figure is obtained by dividing "Cycles number per image processing" by number of computed pixels.
- *Operator load*: Percentage of SIMD processing power used by the operator.

We can get cycles number to obtain 25 Hz rate: $873{,}810 \times 25 = 21{,}845{,}250$ cycles per second. Since SIMD works at 150 MHz SIMD load for this algorithm is 14.56%.

Table 16.3 Processing power at operator and algorithm level

	Number of Operator Calls	Cycles Number Per Image Processing	Cycles for an Image Per Pixel	Operator Load (%)
Abs	1	15,540	0.031	0.26
Binarise	2	31,140	0.062	0.52
Max	1	15,540	0.031	0.26
Add	1	7,950	0.016	0.13
Mul	1	15,630	0.031	0.26
Substract	1	7,950	0.016	0.13
Weighted add	1	8,100	0.016	0.13
Convol3_3abs	2	152,880	0.266	2.55
Dilate	2	412,720	0.718	6.88
Erode	1	206,360	0.359	3.44
	Total	873,810	1.546	14.56

- Design time/cost:
  - The cost to design the phase 1 platform is 24 MM: 18 MM devoted to architecture design and VHDL design, 6 MM devoted to middleware.
  - The cost to build and integrate the demonstration environment is 8 MM.
  - The cost to microcode the operators is 1 MM.
  - The cost to implement the application on the top of this is ½ MM: one week for dimensioning and 1 week for coding and integrating.

Compared to these figures, the cost estimate to implement the same application the traditional way (hardwiring application in VHDL) is 18 MM.

Such a comparison clearly demonstrates the dramatic gains of productivity due to reconfigurable computing especially when coupled with a multi-application platform approach and a powerful development environment.

The figures measured show that the implementation well fits the intermediate platform.

16.4 Phase Two Achievements

16.4.1 Presentation

TOSA'S objectives for phase two are:

- To simulate the test case application on the MORPHEUS chip simulator using the MORPHEUS toolset
- To get measures from this simulation

To extrapolate from these measures some assessment figures of the MORPHEUS chip against the metrics defined.

Contrary to phase one which is a real-time demonstrator, phase two is a simulation and doesn't include a video acquisition module. The input is simply composed of two images, one of the background of the scene and one of the current image of the scene. The output will be one result image. The demonstrator doesn't process sequences but images. Measures are extracted from this simulation and extrapolated so as to be compared against phase one metrics.

At the motion detection algorithm level a little change is made in order to adapt the processing to the need. Accumulation part in phase one which has sense while processing sequences but not images is replaced by subtraction of current image by a reference image defined at the beginning of the processing. Figure 16.4 defines the algorithm that is implemented in phase two.

16.4.2 MORPHEUS Considerations

MORPHEUS chip architecture and TOSA phase one intermediate platform processing engines are similar enough to use the same scheme to map the motion

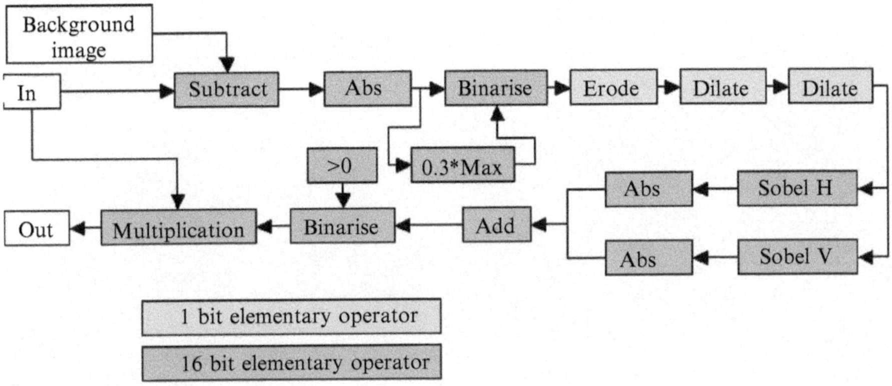

Fig. 16.4 Phase 2 motion detection algorithm

detection application on both devices. In phase one, intermediate platform processing supervisor is the μBlaze. In the MORPHEUS chip it is the ARM. In phase one the bulk processing HRE was the SIMD, in phase 2 the HRE used by TOSA will be the PiCoGA. Since a very interesting work has been done by ARCES and Claudio Mucci to map the motion detection algorithm on the PiCoGA TOSA is taking advantage of this work as a baseline to build phase two simulation.

The HRE choice is driven by estimation of metrics and needs. This estimation has been made during the phase one. The choice is made by the following considerations:

If the requirements (as measured by the metrics, for instance bandwidth, …) are low, HRE choice criteria would be the easiest and fastest way of development. If figures are high, HRE choice criteria would be coding at a lowest level, tuning all the chip possibilities. After examination of the figures concerning our application, it appeared that figures were not so high and our choice has been to choose the PiCoGA which have many advantages:

- Sufficient processing power and memory bandwidth in case of taking advantage of binary processing when possible
- Sufficient memory available to process tiles
- A generic high level programming language (GriffyC)

16.5 Conclusions

From TOSA point of view MORPHEUS seems to fulfill the intelligent surveillance market requirements. A whole implementation methodology has been defined allowing high re-use capability and high productivity. Implementation and simulation shows that MORPHEUS architecture is flexible enough to address general purpose applications. The toolset associated to MORPHEUS chip should provide ease of implementation to a non-expert which is mandatory for a wide spreading of the chip.

Chapter 17
PHY-Layer of 802.16 Mobile Wireless on a Hardware Accelerated SoC

Stylianos Perissakis, Frank Ieromnimon, and Nikolaos S. Voros

Abstract Modern-day wireless systems are characterized by short development windows, coupled with a large volume of requirements that are often antagonistic to each other, such as ever-increasing processing power, placed against the need to keep power consumption from growing, and also the need to improve on the efficiency of use of the RF spectrum. Traditional ASIC design approaches are severely strained especially by the requirements for fast design turn-over and adaptability to a diverse RF environment and alternatives are eagerly searched for in many disciplines. One promising alternative is the use of reconfigurable ASIPs, which offer the promise of speedy development and upgrade of complex systems, combined with the ease of use and relative safety of software-programmable components. MORPHEUS is one such platform that is evaluated in this case study by means of a portion of the PHY-layer for the popular 802.16e/j wireless standards.

Keywords Wireless standards · ASIPS · programmability

17.1 Target System Overview

The application targeted by Intracom Telecom Solutions is part of the emerging IEEE 802.16j standard. The latest standard currently in force from the IEEE 802.16 family [1] is 802.16e, the basis for Mobile WiMAX technology [2]. This standard mandates the use of Orthogonal Frequency Division Multiple Access (OFDMA) technology for the physical layer [3] and provides all necessary support in the physical and MAC layers for mobility management, such as network entry, handover, etc. The next standard, 802.16j [4], currently in preparation, extends

S. Perissakis (✉) and F. Ieromnimon
Intracom Telecom Solutions S.A., Greece
sper@intracom.gr

N.S. Voros
Technological Educational Institute of Mesolonghi, Dept. of Telecommunications Systems & Networks (consultant to Intracom Telecom Solutions S.A.), Greece

N. Voros et al. (eds.), *Dynamic System Reconfiguration in Heterogeneous Platforms*, Lecture Notes in Electrical Engineering 40,

the concepts defined in 16e by adding the possibility of multi-hop communication between mobile and base station. For this, the Relay Station entity is defined. The Relay Station is connected to the base station on one side and to a group of mobile stations on the other. The connection to the base station, where the relay acts more or less as a subscriber/mobile station, is called the “relay link”, while the connection to the mobiles, where the relay acts as a simple base station, is called the “access link”. BS-MS communication now may take place over two hops (BS-RS and RS-MS), which can be advantageous because a poor channel is divided into two better ones, allowing for more spectrally efficient modulation and coding schemes. Alternatively, the range of a network cell can be extended, with relays placed near its periphery, serving distant mobiles. The 802.16j standard will reuse the OFDMA physical layer from. 16e, with some minor enhancements regarding mainly the frame structure, and will make significant amendments to the MAC layer.

Relays can be fixed (located e.g. on rooftops, lamp posts, etc.), nomadic (transportable, e.g. on trucks) or mobile (on buses, trains, etc.). Here we focus on the case of a Mobile Relay Station (MRS). Figure 17.1 demonstrates the concept of a multihop network, including an MRS mounted on a bus that provides service to passengers onboard. As the MRS moves within an area, it will have to perform handover between different base stations (when crossing from one network cell to another). At the same time, the group of mobile stations it supports will also change dynamically over time. The physical layer mode used in each cell is determined by the base station that serves it. As the propagation environment differs from cell to cell (e.g. urban, suburban, rural), different base stations may require different physical layer modes. While simple terminals, supporting only the mandatory modes, are still backward compatible with all base stations, they need to be able to support the advanced modes in order to take advantage of them. The same holds for a MRS, which acts as a terminal on the relay link.

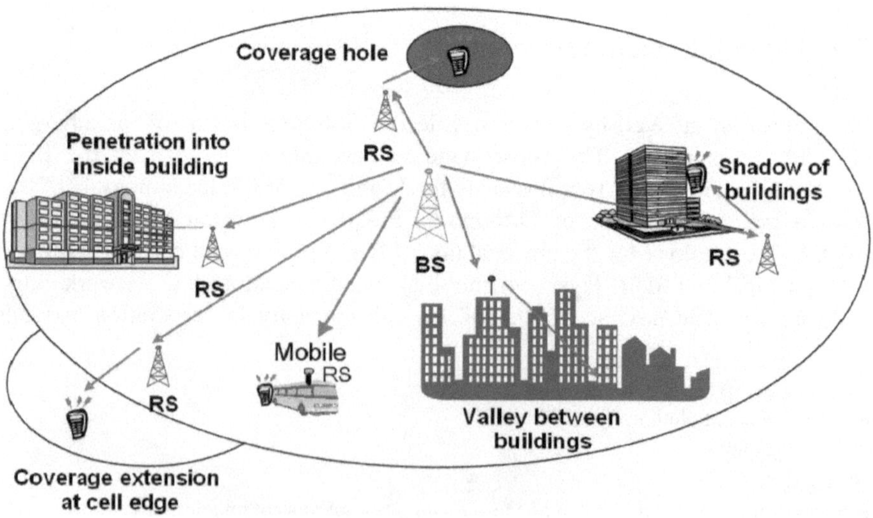

Fig. 17.1 Multihop network concept

The use of a reconfigurable platform for this application is attractive, since it offers:

- Long-term flexibility (the ability to adapt to future amendments of the standard).
- Short-term flexibility (the ability to switch transmission modes dynamically, e.g. during handover). Note that a limited initial selection of transmission modes can be extended in the field with additional modes, so it is desirable not to commit to any specific algorithms at system design time, as would be the case with an ASIC platform.

For the first trial of the MORPHEUS concept, we focused on aspects such as exploitation of available concurrency, as well as evaluation of the efficiency of the algorithm mapping process, i.e. how well do the MORPEHUS HREs accommodate the chosen application components.

17.1.1 Specification of the Envisaged Application

Taking into account the final position on the availability and capacity of the MORPHEUS embedded blocks, Intracom Telecom Solutions performed a study so as to select functional blocks that would (a) fit in these embedded blocks and (b) match the processing nature of these blocks. The result of this line of reasoning was the decision to use the FlexEOS and DREAM HREs described in Chapters 4 and 5, respectively.

A computationally intensive word-level processing block was selected (FFT) for the DREAM section (a 128-point FFT), while a QAM symbol demapper was selected for the FlexEOS section. Therefore the selected functionality focuses on that part of the physical layer of IEEE 802.16j standard that would meet both the above capacity constraints and the processing nature of the embedded blocks.

Additional Mobile Station Receiver blocks are also implemented in MORPHEUS chip, such as Cyclic Prefix Removal and Guard Removal. These blocks are implemented by employing data transfers of specific profiles, involving internal memory. The part of the receiver chain that will be implemented is shown in Fig. 17.2, while the blocks included in the chain are briefly described as follows:

- *Data input*: Continuous stream of complex time-domain data. Logically the stream consists of *Frames*, each frame consisting of a number of *Symbols*, each symbol consisting of 144 complex samples. Depending on frame configuration, we have 1 preamble symbol, a number of downlink data symbols (typically 28), and a number of "garbage" data samples, corresponding to the uplink period.
- *Cyclic Prefix Removal*: Based on control information coming from the Preamble Detection/Synchronisation Block, a number of data samples from the continuous input stream are discarded, and a continuous 128-sample window of data (or complex time-domain symbol) is passed to the FFT downstream.

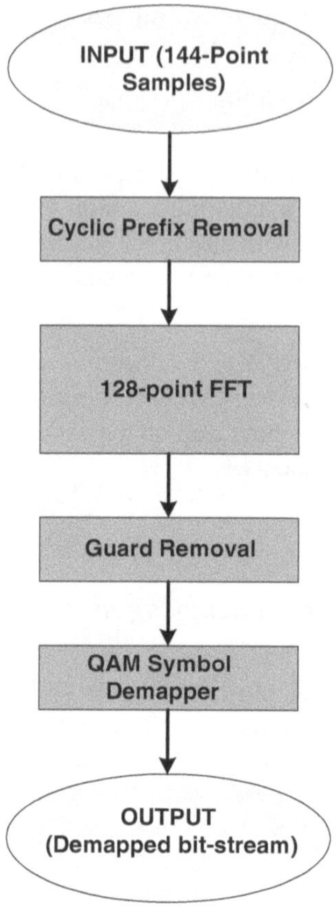

Fig. 17.2 Outline of the proposed wireless application example

- *FFT*: Performs the Fourier Transform on every group of 128 samples, resulting in a complex frequency-domain signal, with 128 points for every input symbol.
- *Guard Removal*: For normal operation this block removes the guard subcarriers from each symbol.
- *QAM Symbol Demapper*: The complex frequency-domain data represent groups of data bits, the number of which depends on the chosen modulation scheme, e.g. 2 bits for QAM4, 4 bits for QAM16, etc. The QAM Demapper converts the points of the complex plane to the groups of data bits that were originally mapped into the constellation of points and converted into a time-domain symbol by the transmitting IFFT.

17.2 Implementation Considerations

As already outlined briefly, the mapping of the various components of Fig. 17.2 was driven by estimates on the capacity of each HRE to handle the computational load. Therefore, the much-larger capacity DREAM array of reconfigurable arithmetic elements was best suited for the FFT block. However, the required buffer space for storing working data was selected to be inside the MORPHEUS chip Local SRAM. The SRAM size of 256Kbytes is enough to cover double-buffering of FFT sample-data. This feature allows speed optimization, as it allows concurrency between data processing by the DREAM ALUs and data transfers between local SRAM buffers and the external memory that represents the MORPHEUS chip environment (i.e. data source & sink).

The task of QAM demapping was delegated to the FlexEOS array. The much smaller size of that particular HRE is a severe constraint that has to be met by the application design. Thus, rather than be prolific with the description of the structures to be rendered in VHDL, we rely again on the fast memory-to memory transfer capabilities furnished by the MORPHEUS chip, to offload some of the "housekeeping" computation onto the ARM9 host.

17.3 The Adopted Strategy

17.3.1 Platform and Tools

The MORPHEUS toolset flow described in Part III of the book has been employed for mapping the application onto MORPHEUS chip. Not all of the C files written for the application are processed by the MORPHEUS toolset flow. Test-data generator/checker routines have also been created alongside the files that describe the wireless components defined in Section 17.1. These generator and checker routines are used for producing the data stream that is supplied to the MORPHEUS chip during the simulation phase. The Modelsim simulation environment is also provided with modules external to the MORPHEUS chip, required for managing the data-stream produced by the generator routines.

17.3.2 System Overview

The wireless application was originally modelled as a C program, running on a PC. The program, shown schematically in Fig. 17.3, consists of a sample stream generator and file-writer function, the application mapping functions, applied to the data in sequence and finally a file-writer function, taking as input the bit-stream generated by the last application module.

The sample stream generator creates a data-stream consisting of 144-sample-point symbols that are forwarded to the Cyclic-Prefix Removal function. The same data-stream is formatted for dumping into a VHDL-readable ASCII file, which is used during the VHDL simulation phase (see Fig. 17.5).

The bit-stream produced by the QAM Symbol Demapper function is dumped by the second file-writer into a VHDL-readable ASCII file. This second file

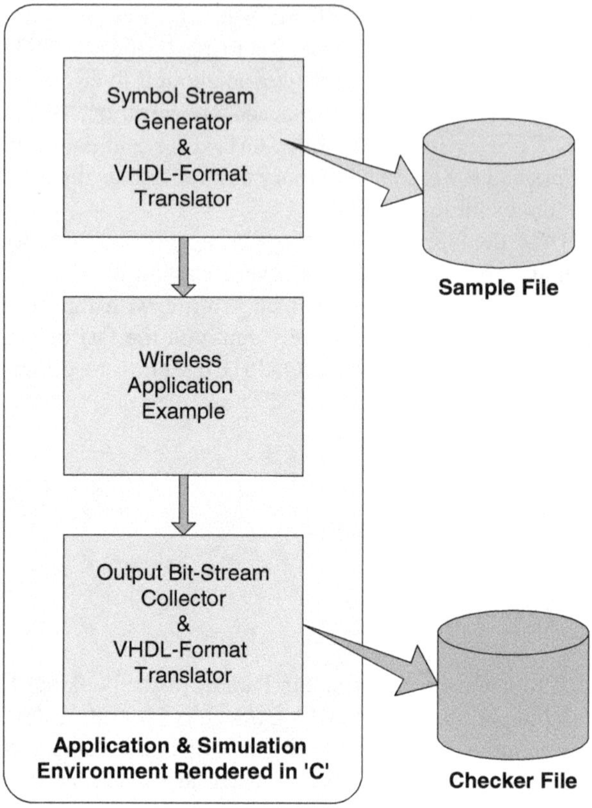

Fig. 17.3 Outline of the application template/test-data generator implemented in C

is employed during the VHDL simulation phase for validating the output of MORPHEUS against the output produced by the reference C simulation.

The C template is subsequently used as the starting point for application of the MORPHEUS toolset, which generate the configuration bitstreams loaded into MORPHEUS. The chosen configuration for the application is shown in Fig. 17.4. Local SRAM is employed for double-buffering of sample data and of the generated bit-stream. The FFT resides in the DREAM array, and two DEBs are employed for communication with the rest of the application. The constellation demapper and guard-removal blocks are inside the FlexEOS array, again employing DEBs for I/O. A final buffer residing in the local SRAM is employed for aligning the generated bit-stream to the external memory's word-length, in order to maximize transfer efficiency. All inter-block transfers are done by DMA channels, programmed and controlled by the ARM9 host.

The executable binary produced by the MORPHEUS toolset is converted for use in a VHDL simulation environment, shown in Fig. 17.5. This environment consists of the MORPHEUS modules representing the chip, attached via the external

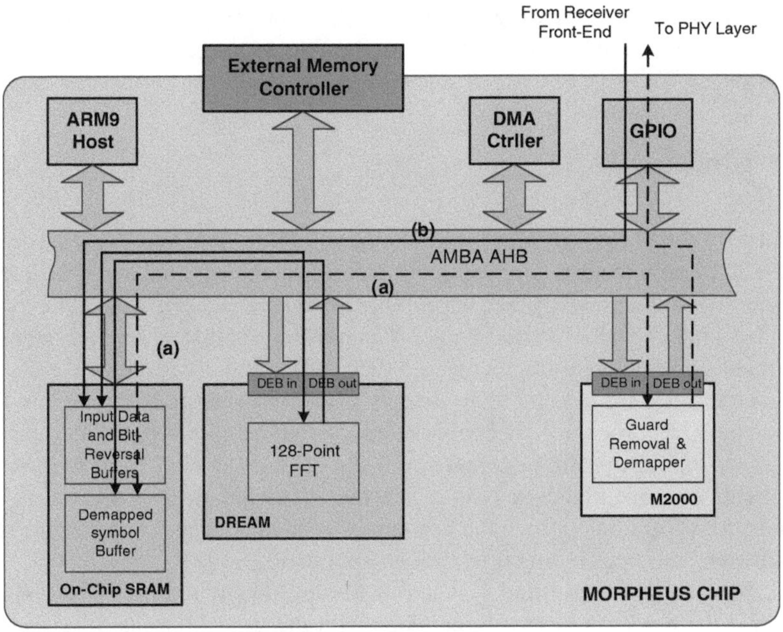

Fig. 17.4 Illustration of the application mapping onto MORPHEUS

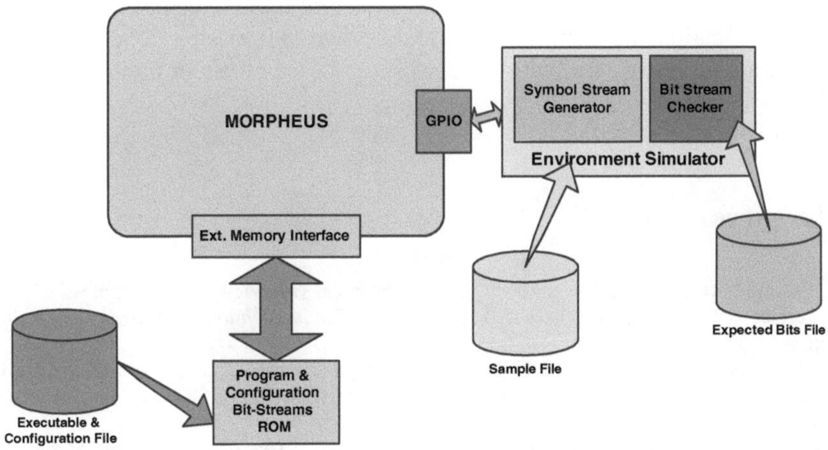

Fig. 17.5 The simulation environment for the wireless PHY segment on MORPHEUS

memory interface to a "ROM" model holding the executable image that is loaded and run by the ARM9 host of MORPEHUS chip. A data-stream driver is attached to the MORPHEUS GPIO, sending the translated patterns that were saved by the C simulator data generator/file-writer. A bit-patterns checker is attached to the

GPIO's output lines, comparing the MORPHEUS output with the contents of the file that was generated by the C application bit-stream logger.

17.4 Conclusions

A small part of the PHY layer for the 802.16e/j wireless standard was ported onto the MORPHEUS architecture, a novel combination of conventional software-running processor with dynamically reconfigurable hardware accelerators. The porting was aided with the SPEAR and Molen Compiler tools that allow the capture of performance-critical code fragments and their replacement with library calls to an API that translates the passing of arguments to data-processing software into DMA transfers between processor-accessible memory and the hardware accelerators represented by the API primitives. Synchronization issues arising from the mapping of segments of sequential code onto concurrent hardware are also taken care of by the tools. An RTOS onto which the user application is attached manages the use of the hardware resources by the transformed application code.

A simulation of the resulting system in a ModelSim environment was employed to validate the porting process. Experience with the use of the platform and tools showed that the approach is promising. The tool-chain front-end is of particular importance from the user's point-of-view, and is seen as the most dynamic aspect in the area of tool development for platforms such as MORPHEUS. This is because most of the issues related to exploitation of concurrency and reconfigurability arise in the specification and capture phase. Thus, progress in this part of the tool chain can have a multiplicative effect on the scope and ease of use of platforms such as MORPHEUS.

References

1. C. Eklund, et al., IEEE Standard 802.16: A Technical Overview of the WirelessMAN Air Interface for Broadband Wireless Access, *IEEE Communications Magazine*, June 2002, pp. 98–107.
2. WiMAX Forum, Mobile WiMAX – Part I: A Technical Overview and Performance Evaluation, February 2006. http://www.wimaxforum.org/sites/wimaxforum.org/files/documentation/2009/mobile_wimax_part1_overview_and_performance.pdf.
3. H. Yaghoobi, Scalable OFDMA Physical Layer in IEEE 802.16 WirelessMAN, *Intel Technology Journal*, 8(3), August 2004, pp. 201–212.
4. IEEE Standard for Local and Metropolitan Area Networks – Part 16: Air Interface for Fixed and Mobile Broadband Wireless Access Systems – Draft Amendment j: Multihop Relay Specification, IEEE P802.16j/D2, December 24, 2007.

Part V
Concluding Section

Chapter 18
Conclusions

MORPHEUS Reconfigurable Platform – Results and Perspectives

Philippe Bonnot, Arnaud Grasset, Philippe Millet, Fabio Campi, Davide Rossi, Alberto Rosti, Wolfram Putzke-Röming, Nikolaos S. Voros, Michael Hübner, Sophie Oriol, and Hélène Gros

Abstract The MORPHEUS architecture principle, plus its associate toolset, bring together a significant advantage for embedded system designs: performance, flexibility and productivity. The project also prepares, to a certain extent, the future utilization of reconfigurable technologies complementarily to multi/many-core solutions.

Keywords Reconfigurable computing • SOC • heterogeneous architecture • performance density • execution flexibility • programming productivity • project consortium

18.1 MORPHEUS Project

The dynamically reconfigurable SOC approach presented in this book has been studied and developed in the frame of the MORPHEUS European project under the 6th Framework Program. MORPHEUS started in January 2006 with main objectives

P. Bonnot (✉), A. Grasset, and P. Millet
Thales Research & Technology, France
philippe.bonnot@thalesgroup.com

F. Campi, D. Rossi, and A. Rosti
STMicroelectronics, Italy

W. Putzke-Röming
Deutsche Thomson OHG, Germany

N.S. Voros
Technological Educational Institute of Mesolonghi, Dept. of Telecommunications Systems & Networks (consultant to Intracom Telecom Solutions S.A.), Greece

M. Hubner
ITIV University of Karlsruhe (TH), Germany

S. Oriol and H. Gros
ARTTIC SAS, France

N. Voros et al. (eds.), *Dynamic System Reconfiguration in Heterogeneous Platforms*, Lecture Notes in Electrical Engineering 40,

(corresponding to the needs described in the introduction chapter) to improve performance density, reconfiguration flexibility and programming productivity.

The project consortium grouped a number of partners involved in studying and providing the required technologies or specifying the requirements and assessing the results.

The MORPHEUS consortium consisted of:

- Thales Research & Technology (TRT, France),
- Deutsche THOMSON OHG (DTO, Germany),
- INTRACOM SA Telecom Solutions (ICOM, Greece)
- ALCATEL-LUCENT Deutschland AG (ALU, Germany)
- Thales Optronics SA (TOSA, France)
- STMicroelectronics SRL (ST, Italy)
- PACT XPP Technologies AG (PACT, Germany)
- M2000 (M2000, France)
- ACE Associated Compiler Experts bv (ACE, The Netherlands)
- CriticalBlue (CBlue, United Kingdom)
- Universitaet Karlsruhe (UK, Germany)
- Technische Universiteit Delft (TUD, The Netherlands)
- Commissariat à l'Energie Atomique (CEA, France)
- Université de Bretagne Occidentale (UBO, France)
- Universita di Bologna (ARCES, Italy)
- ARTTIC SAS (ARTTIC, France)
- Technische Universitaet Braunschweig (TUBS, Germany)
- Technische Universitaet Chemnitz (TUC, Germany)

The Executive Board of the project (see Chapter 22 on project management) was composed of TRT, ARTTIC, DTO, ICOM, ST and UK.

The coordination of the project was ensured by Thales Research & Technology. ARTTIC, in charge of day-to-day management and administrative aspects, efficiently handled with the difficulties encountered during the project (new partners, budgets reallocations, planning modification, duration extension and process for important decisions).

The main outcomes of the project are the definition of the whole hardware/software approach for dynamic reconfigurable SOC, the circuit fabricated in ST 90 nm technology and the toolset allowing a seamless flow from high level description to the implementation of applications. The following sections summarize these achievements.

18.2 Architecture Achievements

The MORPHEUS architecture chapters (Chapter 3 and following) explained the approach of using an ARM as central controller to manage the whole platform and the usage of the Molen paradigm to control the HREs from this central processor. It also explained the different data processing models supported by the architecture: the data-stream processing principle corresponding to the execution of a pipeline across multiple HREs on one hand, and the repeated usage of same HRE for consecutive processing steps involving thus the reconfiguration of the HRE on the other hand.

These baseline principles are completed with architecture features enabling modularity and scalability of the platform. The encapsulation of HRE thanks to DEBs, CEBs and XRs leverages the exchange of certain HREs with another one from the set of available HREs. The MORPHEUS platform is indeed an architectural framework that only defines how modules can be integrated rather than proposing the usage of specific HREs. For a given application domain, a specific architecture must be derived from this platform. The concept of the MORPHEUS approach was thus evaluated on a specific demonstration architecture derived from the architecture principles, and implemented on the MORPHEUS prototype chip produced during the project.

The architecture chapters also highlighted the advantages of the memory hierarchy and the communication means aiming at providing the computational engines with the necessary data throughput while retaining ease of programmability. This includes first a computation model capable to hide heterogeneity and hardware details while providing a consistent interface to the end user. This is also complemented by the data storage and communication infrastructure adapted to all the different flows defined over the architecture in its lifetime. This concerns data as well as configuration flows and involved Network-on-Chip (from ST, Chapter 8) and Data Network Access mechanism (from Universitaet Kalsruhe, Chapter 8) for high internal bandwidth and scalability motivations, Predictive Configuration Manager (from CEA, Chapter 7) for efficient dynamic reconfiguration management. Even if the DDR memory controller (from TUBS, Chapter 7) was finally not inserted in the fabricated version of the chip, it illustrates the important attention given to data bandwidth for such kind of approach.

Naturally the advantages of the different HRE technologies including eFPGA (from M2000, Chapter 4), DREAM/PICOGA (from ARCES and ST, Chapter 5) and XPP (from PACT, Chapter 6) are key elements for the whole system.

The chip size is 110 mm^2, including pads. HREs occupy around 64% of the overall area, the rest being divided between processor, communication infrastructure and 256 pads. The overall chip contains 8,500 equivalent standard cell Kgates, 8,831Mbits of embedded memory cuts and a total number 97 millions of transistors. XPP can run at 150 MHz and DREAM/PICOGA at 180 MHz.

Dynamic power estimation is roughly 3 W for a working frequency of 100 MHz for the HREs and 200 MHz for the control and communication system.

18.3 ToolSet Achievements

Because of its heterogeneity and the necessity to handle dynamic reconfigurations and efficient data communications, the MORPHEUS platform architecture is quite complex. The Chapter 9 on toolset overview, as well as the complementary chapters on programming topics, explained that this complex architecture especially requires a global and seamless tool chain. Such tool chain has therefore been developed. It starts from high level description language down to executable code for the hardware platform. The capabilities of this tool chain include the programming of the system which encompasses the management of global control, synchronization, reconfigurations and communications. They also include the capability to generate the configuration code for the various HRE.

The programmability easiness at global level is brought by two main elements that are the MOLEN concept (from TUD, Chapter 10) and the ISCR library (from Universitaet Karlsruhe, Chapter 11). MOLEN, on one hand, indeed permits to easily deal with management of accelerated functions at compilation time within CoSy compiler (from ACE, Chapter 10). Instead of dealing with all the burden of preparing the configuration, launching the communication DMA and the execution of the function itself, the programmer only has to insert a pragma to specify that the function must be accelerated. The ISRC library of RTOS services for dynamic reconfiguration management, on the other hand, allows the control at runtime of the acceleration driven by the MOLEN directives. The ISRC library is completed with dynamic memory management services (from ICOM). The MOLEN directive advantage is also to comply with the OpenMP directives, especially for the management of parallel threads on the system.

Concerning the generation of HRE configuration code the challenge was also to address the design of time and space implementations. The first step of the approach consists in capturing the description of the function with all its inherent parallelism. This is made possible for the addressed application cases thanks to the SPEAR tool (from Thales TRT, Chapter 12) to describe a chain of interconnected kernels handling with multidimensional data arrays, thanks to CASCADE (from CriticalBlue, Chapter 12) to generate kernel code from C language code and thanks to SPECEDIT (Alcatel-Lucent tool, Chapter 12) for the capture of finite state machines with formal checking capabilities. These SPEAR, CASCADE and SPECEDIT tools are capable to generate Control Data Flow Graphs from which the MADEO framework (from UBO, Chapter 13) can synthesize either a netlist for eFPGA HRE or the code for the DREAM HRE.

18.4 Relevance of the MORPHEUS Approach for the Selected Applications

The various applications which brought the main requirements for the MORPHEUS platform benefit from the approach with different perspectives.

They all naturally make use of the acceleration made possible thanks to the HRE (and generally several of them). They also all take advantage from dynamic reconfiguration, but with different justifications.

For the video image de-noising system (from DTO and TUBS, Chapter 14), the intensive data-stream aspect is prominent. The utilization of dynamic reconfiguration of the XPP HRE (several times per frame) is justified by the necessary reutilization of that accelerator for several functions that must be subsequently computed. This corresponds to a clear motivation for performance efficiency.

For the networking application (from ALU and TUC, Chapter 15), the dynamic reconfiguration of the M2000 eFPGA HRE is mainly justified by the need to update the behavior of the system. Reconfiguration is indeed performed according to "reconfiguration packets" corresponding to network updating information.

For the motion detection algorithm of the video surveillance smart camera (from TOSA, Chapter 16), the usage of dynamic reconfiguration is also driven by the philosophy of implementation which aims at optimizing the performance density by reusing the HRE for different functions.

For the wireless application (from ICOM, Chapter 17), the reconfiguration is mainly justified by the ability requirement to switch transmission modes dynamically.

18.5 MORPHEUS Platform Achievements and Future Steps

Consequently to the achievements presented throughout the chapters of this book in terms of architecture and programming solutions, the MORPHEUS project claims that the proposed approach permits to reach the essential objectives identified for embedded systems development.

The proposed architecture principles indeed provide the expected processing performance (with respect to area and power consumption) because the reconfigurable accelerators are themselves efficient for the type of processing they address and also especially because of the dynamic reconfiguration which allows the powerful reutilization of the same hardware when necessary and finally because of the efficient communication system. In spite of the constraints encountered during the project, the chip permitted to assess this efficiency on the target applications.

Regarding the aimed flexibility, here again the reconfigurable technologies bring themselves this advantage. Coarse grain XPP, multi-context PICOGA or small eFPGA block can be reconfigured quickly enough to enable the dynamic reconfiguration within the system. The RTOS services that have been especially developed and reinforced by the PCM hardware module allow an efficient usage of this dynamic reconfiguration.

Moreover, the crucial objective of a high application implementation productivity is enabled by the overall toolset thanks to the way it deals with offloading the reconfigurable accelerations and generating configuration code from C language programming plus relevant complementary information that has to be provided in a simple manner (pragma directives and graphical interface).

Globally, it has thus been claimed here that the MORPHEUS architecture principle plus its associate toolset bring a significant advantage for embedded system designs.

What can be the next steps for this type of SOC approach in the future? The proposed system-level principles can certainly be extended and enhanced (larger number of accelerators, more powerful allocation schemes, higher abstraction programming level, more efficient memory interfaces, etc.). The crucial point will probably be the identification of new embedded reconfigurable technologies.

Current offers are not numerous and still address a quite small market. This might change, especially when the habits will change, when users will better understand how much these technologies can be an alternative or a complementary solution to multi/many-core solutions.

For this raison, training is also important in the MORPHEUS approach. First, even if everything was done in the project to make the platform easily handled by programmers only used to classical software programming, the usage of the platform can be improved if the programmer is aware of the advantages of dynamic reconfiguration and how to handle with it. Generally speaking, training (at university or in industry) on the awareness of combined software and hardware knowledge is essential and is also one goal of MORPHEUS project. This aspect will be presented in the first following chapter.

The following part indeed presents complementary topics related to project aspects: training, as just mentioned above, but also dissemination and exploitation of the project work, and finally management aspects.

Chapter 19
Training

Michael Hübner, Jürgen Becker, Matthias Kühnle, and Florian Thoma

Abstract This chapter describes the training activities within the MOPHEUS project

Keywords Training • courses • requirements • hardware software codesign

19.1 Introduction

System design of the future including both software and hardware components to fulfill the requirements of real-time applications are the focus in the MORPHEUS project. The combination of control flow oriented architectures like microprocessors and parallel processing architectures like FPGA enlarge the design space for system designers and is a challenging task for the provided design flow [1,2]. To ensure the optimized exploitation of the targeted heterogeneous architecture in MORPHEUS, an adequate training must be provided from the academic partners focused to the requirements of the industrial partners. Therefore an assessment of the industry needs was done in order to find gaps and increments in existing courses for undergraduate students but also for well established engineers. The well established courses for software engineers have to be extended in that way, that the difference to the traditional von Neumann notion will be introduced in an early phase of study. The completely new paradigm of system design with reconfigurable hardware, enabling the computation in time and space, has to be integrated in existing courses, but also new course material and lectures has to be provided in order to make the grade for future.

M. Hübner (✉), J. Becker, M. Kühnle, and F. Thoma
ITIV, University of Karlsruhe (TH), Germany
Michael.Hübner@itiv.uni-karlsruhe.de

N. Voros et al. (eds.), *Dynamic System Reconfiguration in Heterogeneous Platforms*,
Lecture Notes in Electrical Engineering 40,

19.2 Description of Training Requirements from Industry Point of View

An important task for the training workpackage within the MORPHEUS project, was to collect the requirements to skills of future engineers from the industrial partners. Especially the novel approach where reconfigurable computing is exploited in a multicore chip forces to consider about new skills for the developers of future electronic systems. The following sections describe the demands from the different industrial partner.

19.2.1 Thales

Thales organized several internal workshops on reconfigurable computing [3,4]. They revealed that relatively few people are sensitive to the question, even if the use of reconfigurable components is widely accepted, but not at the right level. Reconfigurable components are indeed not yet considered as real computing platforms in spite of the fact that they more and more play that role. Thales intends to organize another workshop in order to make people aware of the subject and also to get a feeling of the evolution of the sensitivity within Thales.

This is also the illustration that reconfigurable computing implies a multidisciplinary approach encompassing architecture, SW and HW engineering expertises. Thales considers naturally that two types of engineers are concerned by such training:

- The engineers currently developing hardware like FPGA (it concerns approximately 200 engineers in a company like Thales).
- The engineers currently developing embedded software solutions. Among them, we can imagine that approximately the same amount of persons should be concerned.

Both types of engineers would require different training adapted to their habits and to the type of skills they would have to make use of.

- FPGA designers should be sensitized to the new possibilities offered by a platform like MORPHEUS. Concerning the architecture, they should be made aware of the computation power of coarse-grain modules, the simplification brought by the memory hierarchy, the efficiency brought by the on-chip network. Concerning the toolset, they should be made aware of the flexibility offered by the dynamic reconfiguration management and the design productivity made possible thanks to tool modules offered to implement functions on reconfigurable units. Finally they should get the skill corresponding to the possibilities offered by MORPHEUS and not offered by FPGA. These skills will be detailed by academic partners of the project.
- Software designers should be made aware of the speed-up brought by the Molen paradigm making use of the reconfigurable units offered by the architecture, the easiness of implementing functions on reconfigurable units to be able to evaluate the interest of such solution.

All of them should follow training on the usage of the MORPHEUS platform and the different technologies involved inside: PACT XPP, ARCES PicoGA, M2000 Flexeos architectures and tools, Molen compilation, Dynamic reconfiguration RTOS, RC-SPEAR and MADEO design tools, etc. Expected pilot demonstrators used for such educational purposes should thus be found among Delft Molen demonstrator, Karlsruhe partial reconfiguration demonstrator, PACT demonstrator, PicoGA demonstrator, etc.

Potential friendly industrial user interested in such training could be find among companies working on embedded computing products in domains of communications, multimedia, automotive, security, etc. and also among partners of current or past IST projects on reconfigurable and/or advanced computing like Shapes, Sarc, Aether, 4S, Reconf.

But, the global objective of this type of training should also be to produce a new sort of engineers (either just coming out from university or more experimented ones) aware of both software and hardware issues and convinced of the interest of the good flexibility/performance compromise solution that is offered by reconfigurable computing. This means for example: making software designers sensitive to reconfigurable computing in general, giving them another vision than the classical von Neumann paradigm they have been using for ages without being conscious of doing so (giving them for example the vision of more dataflow oriented machines instead of control flow oriented machines, etc.). This means also make engineers aware of the panorama of reconfigurable architecture existing both on market and academic studies (Twente Montium, IMEC ADRES, Elixent DFA, Stretch ISEF, etc.) and the interest of generic technologies like dynamic reconfiguration, etc. also providing them with information on the maturity of such technologies.

19.2.2 PACT

PACT is aware of the fact that future products will be defined mainly by software. The underlying hardware layer is not more than a vehicle to implement the application. This impacts the needs for training courses.

19.2.2.1 Training for Students

Looking at the tremendous complexity of video and wireless applications, it is nearly impossible to develop and port the software for a complex SoC such as the MORPHEUS chip without using Software IP from external suppliers. Nevertheless, the integration of the externally developed modules (hardware and software) is the responsibility of the hardware suppliers. This requires additional skills not only for graduate students from the universities but also for experienced engineers. The required skills are moving towards “system engineering”.

The following list summarizes the topics which should be addressed by training courses which prepare students to deal with new architectures.

Software Designers …

- **Must understand new hardware architectures.** They must be aware of the hardware structures on which their software will be executed. This requires profound understanding of the mostly very complex chip structures and the hardware capabilities. One methodology for new software designs is to specify modular structure which allows a step-by-step porting from sequential microprocessors to the target architecture. This is especially important for reconfigurable architectures which allow a very high potential for optimization of performance critical code sections. To be trained: hardware understanding on system level and step by step mapping to the architecture.
- **Must be able and willing to leave high abstraction levels.** Not optimized software requires too many hardware resources or cannot perform the task on a given hardware platform. Thus, the software must be profiled and critical sections need to be optimized either by assembler or using native tools for the available reconfigurable hardware. To be trained: profiling and native programming.
- **Must have an early understanding of required communication bandwidth.** Reconfigurable hardware structures are able to process very high bandwidth data streams. The bottleneck is often the memory and data exchange. Therefore intelligent handling of local buffers and usage of the memory hierarchy is mandatory. Since often the final hardware is not available at the time when the software is specified, estimation and simulation of the data transfers should be done at an early stage. In many cases evaluation and estimation can be done with dummy functions. To be trained: bandwidth estimation and simulation on system level.
- **Must know how to handle distributed multi-tasking systems.** Synchronization of tasks which are executed on hardware structures is important. Related is the efficient handling of data buffers. To be trained: Multi tasking and synchronization of tasks and data transfers.
- **Should specify software-hardware interfaces.** Interfaces must be specified with the goal that driver software can be written efficiently and with highest performance. To be trained: How to write efficient drivers and how to specify the hardware for it.
- **Should know how to specify external IP.** Many sub-tasks of new designs must be outsourced to external suppliers or other working groups. This can only be successful if the tasks to be outsourced are specified unambiguously with simple interfacing. The test and acceptance procedure must be specified too. To be trained: structure of specifications and level of details and acceptance procedures.
- **Should be able to handle projects.** Project management is not (only) to prepare Gantt charts, it is to access continuously the project status and risks and to ensure the communication between the involved groups. A software project manger must be aware that all contributions fit together according to the specified interfaces. To be trained: how to avoid iteration "loops" because of misunderstandings or unclear specifications, how to check the plausibility of time estimations and how to steer complex projects.

Hardware Designers …

- **Must be aware that software defines the product at the end.** Thus, hardware designers must request at an early design stage the application and software requirements. On the other hand, it is important to evaluate the "costs" (area, power) of software demands. To be trained: Translate software demands into hardware requirements and calculate the "costs" of those requirements.
- **Must design to test.** Verification of complex SoCs is becoming a critical phase in chip designs. Thus hardware must be modular so that modules can be tested standalone as far as possible without the need for a complete system simulation. To be trained: modular and testable hardware designs and specification of test suites.
- **Should know how to map sequentially specified algorithms to parallel structures.** Mostly time-critical algorithms are specified sequentially (e.g. nested loops). Hardware engineers must be able to translate this into parallel structures e.g. to data flow graphs which can be mapped on parallel hardware structures. To be trained: Mapping of algorithms to parallel arrays of processing elements (course and fine grained).
- **Must cooperate with software designers.** A hardware–software interface that can be programmed efficiently is mandatory. To be trained: How efficient driver software is designed and which hardware prerequisites are required.

Beside these concrete points to be trained, it is important that both, hardware designers and software designers are aware of "thinking" of the other "domain". Involved people must know that a successful result or product can only be achieved if both domains cooperate perfectly not only on technical level but also on interpersonal relationships.

19.2.2.2 Training Specifically for the XPP Architecture

PACTs strategy is to provide tutorials and examples for the tools and the XPP-III architecture. PACT rather spends efforts to generate examples and documentation on all complexity levels than provides personal for continuous support of novices and customers.

19.2.2.3 Additionally Required Training Material

Based on the XPP tutorials, additional training material will be prepared by the universities extending the tutorials by:

- Methods for partitioning of algorithms
- Methods for optimal memory usage in a heterogeneous architecture
- Examples for partitioned algorithms for the XPP array and FNC-PAEs in the context of the MORPHEUS architecture

- Task synchronisation based on data-flow, Interrupts, XPP event handling and software methods such as Mutex variables
- Examples for DMA setup and buffer handling
- Usage of the APIs and communication features which are developed in the course of MORPHEUS project
- Optimisation methodologies with respect to power and performance

19.2.2.4 PACT Specific Training for Students

PACT provides affordable licenses of its tools and the simulation models for Universities. Thus, the tools can be used in practical training courses in the universities. On the other direction PACT needs feedback to improve its training material.

19.2.2.5 Training on the Job

Theory is good, practice is better. Therefore new engineers are given the chance to go through the new tutorials and learn the XPP by doing under supervision of experienced experts.

19.2.2.6 Training for Customers

PACT provides tailored training plans for interested customers starting from a one day introduction to several days training based on the new training material.

19.2.2.7 Summary

For PACT, comprehensive training material is absolutely mandatory. Without well designed training concepts, there is the big risk, that the new methodologies which are required for heterogeneous reconfigurable architectures will not be accepted by the software developer community [5].

19.2.3 CEA List

CEA LIST has been involved in reconfigurable computing for several years, in projects such as partitioning SW/HW, design of IP, dynamic execution control management for reconfigurable system, dynamic management of resources for reconfigurable system …

MORPHEUS is a great opportunity to promote several concepts such as dynamic reconfiguration (towards the RTOS and the configuration manager), multi task application (allowed by the IPs and the RTOS), multi computation grain system (according to the IPs) and new programming and execution model (according to the configurable memory hierarchy, the RTOS and the IPs). So software engineers and hardware engineers should be aware of these new kinds of system where they are both involved to take the best advantages of the trade off given by the architecture.

The MORPHEUS SOC is very flexible thanks to its plurality of reconfiguration grain and its complex control system. This flexibility enables, on one hand the software engineers to be aware of new software design methodologies for such system and on the other hand, to be capable to take advantage of the hardware.

Applications becoming more and more complex, some decisions such as task scheduling or allocation are dynamically taken in the SOC MORPHEUS. The engineers have to be aware of these mechanisms to take advantage on it. According to the improvement of the architecture, the software designers should be sensitive to new requirements of complex application and design methodology such as multi tasking and dynamic management of hardware resources. For the first point, some courses should be proposed on the design methodology integrating the new requirements of the applications (more dynamic control).

Because some tasks will be mapped on spatial computation engine, students have to understand the trade off that can be done in hardware implementation (performance, surface and power). They should also be aware of the cost of reconfiguration to take the best advantage of this mechanism.

Partitioning of an application will be a key point of the development of application for the MORPHEUS SOC. Students should have some courses to be able to estimate the performance, the power and the surface overheads of several implementations of a task.

On the hardware size, to propose pragmatic architectures and take advantage of new technology, students have to be aware of advanced technology design methodology. On the other hand, because SOC becomes more and more complex, new methodology of conception appears. Students should be especially aware of high level of modelling.

In the MORPHEUS project, the CEA LIST designs a hardware module (configuration manager) involved in dynamic reconfiguration mechanisms. The results of this module will be published and could be used in courses as a case study for dynamic management of reconfigurable resources in SOC.

19.2.4 Critical Blue

CriticalBlue's expertise is in the development of EDA tools for the automatic translation of embedded software into customized coprocessor acceleration hardware. We believe that the main training requirement is in the area of making the MORPHEUS platform efficiently target-able from the perspective of the software engineer.

In particular there is a requirement to cover the typical flow that is required for bringing a software application into a state where it can make full use of the platform. In broad terms these steps would comprise of:

- An initial profiling phase to identify which areas of the application are taking a significant percentage of the runtime and are thus potential candidates for hardware acceleration. An associated and crucial aspect is how the stimulus datasets are designed that drives the application when performing the profiling. This is because in complex algorithms the control paths taken are data dependent and thus the nature of the profile may be significantly affected by the data. The dataset must be representative of the typical behavior but also, if there are real time processing requirements, also cover corner cases. In general the designer must know how to gain an understanding of the sensitivity of the profile to the datasets, otherwise partitioning decisions made at this stage might adversely affect the performance variability of the final system.
- The software engineer must be able to recognize good candidates for hardware acceleration. Clearly a primary criteria is the importance in the profile, but other important aspects include the complexity of transforming implementation into a form suitable for one of the reconfiguration targets on the platform and the communication overhead of the function. The training needs to discuss the typical attributes of functions that are highly amenable to hardware acceleration, i.e. they are data rather than control flow centric in nature and contain significant implicit parallelism. In circumstances where this does not seem to be the case the engineer should gain an appreciation whether this is a fundamental aspect of that part of the algorithm or whether the suitability is actually being obfuscated by an unsuitable coding style that could in fact be transformed.
- The engineer must have training on means to access the likely communication overhead of implementing a particular function using hardware acceleration. This will be determined by how frequently the function is invoked and, for each invocation, how much data need be transmitted into and out of the invocation. Communication can very easily become a bottleneck that completely undermines the benefits of hardware acceleration so a good understanding of this is crucial to the effective use of the platform. The training must cover the communication mechanisms that are used and how well any individual data structure or data layout maps to those communication mechanisms. Ideally the engineer should be equipped with knowledge of how to transform particular data layouts to make most efficient use of the communication mechanisms available. This is a complex area that can involve quite complex code transformations in some cases, but can have a very significant impact on the efficiency of the final implementation. Indeed this area is one of the significant differences between traditional software design and algorithmic description for implementation in hardware, even at a high level of abstraction. Traditional software design typically ignores the overhead of memory access, treating the entire address space as one unified area with identical access cost, whereas any hardware

implementation must treat the overhead of data access as being a significant component of the overall performance. Training materials which illustrate this distinction would significantly aid the targeting of the platform.

- Finally, once the implementation on the platform has been made there must be some real performance analysis to verify the system is running as expected under the required workloads. Where there are significant discrepancies training must be provided for the platform analysis tools so that results may be interpreted correctly and bottlenecks identified.

19.2.5 M2000

M2000's experience shows that the use of field programmable gate array (FPGA) blocks in ASIC/SoC/MCU devices to add 'in-the-field' logic configurability, as well as hardware configurable instruction sets, is not well known among hardware designers, and even less among software designers [6].

For efficient use of the embedded FPGA blocks in an ASIC/SoC/MCU, and to facilitate the adoption of the full set of innovative capabilities offered by the MORPHEUS architecture, M2000 stresses that designers first need a good understanding of the following areas:

- The basic concept of FPGA architecture and the functionalities provided to designers
- The methodology and the mode of use, for both hardware and software

Once the basic concept of an FPGA is understood, the next step would be to appreciate the benefits that reconfigurability brings to computing applications, with an obvious focus on the MORPHEUS architecture.

The benefits of reconfigurability should be illustrated with applications where the differences with and without reconfigurability clearly appear.

A further key aspect is the user software environment and methodology training: how the different tools work together to automate the use of all MORPHEUS functionalities, with particular emphasis on areas that are new, in particular the reconfigurability in its various forms, and how to exploit it to maximize the performance of this new MCU architecture.

To summarize, M2000 considers it very important to train both hardware and software engineers:

- Tutorial on FPGAs: architecture, functionality, modes of use
- Methodology training for MORPHEUS as a whole, with a focus on exploiting reconfigurability
- Workshop: hands-on experience of the complete process using a "real" application example

19.2.6 ST

ST is planning to demonstrate the results from the MORPHEUS project at its internal yearly UNICAD Workshops events in summer 2007 and 2008. Those events will play the role to raise internal interest for applications based on reconfigurable computing, also showing the benefits related to a design flow based on reconfigurable computing.

ST is also evaluating the possibility to include reconfigurable computation as a new course in its STU (ST University) program [7]. After performing training needs analysis within the company, that motivates the need for introducing a new course about reconfigurable computing, the process of including a new training envisages a standard way of presenting the documentation and a certification procedure for the trainer, which can be either internal or external to the company.

ST will propose, in a first phase, a three-day course (24 hours) which will be based on XiRISC-PiCoGA-GriffyC including practical examples, as the course proposed by ARCES. ST will then take into account the possibility to expand this training to a five-day course, including also the treatment of XPP and M2000; the development of this part of the course will depend mainly on the practical possibility to propose to the trainees the same environment that ST uses while developing its architecture, including the needed licenses.

19.2.7 Lucent

Lucent Technologie's goal in the MORPHEUS project is to embed FPGA cores in a telecommunication ASIC to be able to react to IEEE standard updates at late design stages or in the field.

Since Lucent Technologies has a long tradition of hardware development including ASIC and FPGA based techniques, it can be assumed that sufficient knowledge and skills in both of these areas are available in the existing Lucent engineering community.

Graduates and other engineers which never have dealt with ASIC and FPGA development issues before must be educated to fit in Lucent's IC-design world. Following mandatory skills must be present or have to be adopted:

- VHDL coding on behavioural and RTL level, since VHDL is the basic HDL being utilized at Lucent Technologies
- Basic Verilog skills in order to be able to cope with net lists
- Handling and programming of design tools like simulators (modelsim, ncsim), synthesis tools (Synopsys, Synplify) and analysis tools (STA tools, model checker tools)
- Usage of the simulation environment ProVerify, which is a Lucent in-house development
- Understanding of telecommunication standards

Most of these topics are typical demands, which IC design candidates have to meet. Apart from the latter two points training institutions like e.g. universities can cover these demands.

The combination of ASIC and FPGA techniques is a paradigm shift to a certain extend which affects old school IC design engineers also. The effects of the embedding approach reach not only the hardware implementation itself, but also specification and verification phase and finally software development.

In the specification phase a partitioning of static and reconfigurable functionality must be defined. Several constraints have to be considered:

- Which functionality is likely to be changed in the future?
- How to define the interfaces between hard coded ASIC parts and soft coded FPGA parts to be as flexible as possible?
- Which circuit parts can be realized in the FPGA core at all due to technical limitations?
- How to dimension the size of the reconfigurable part?
- How to achieve the requirement that the reconfiguration process must be hitless (i.e. transmission of data is not disrupted by reconfiguration)?
- Considerations to balance financial costs of reconfiguration ability vs. cost benefits coming from update potential.
- Hardware implementation faces new challenges concerning complexity, timing constraints, handling …
- Verification must cover all reconfiguration scenarios. Depending on the application this can multiply the verification effort.
- Software has to be aware, which reconfiguration part currently is active e.g. in order to cope with configuration dependent register maps.

Most of the items mentioned above are objects to investigations and gain in experience. There are no training courses available currently. Thus only the following concrete training areas which are directly connected to the MORPHEUS project have been identified:

- Specific training on the basic technical topic, namely "how to insert an FPGA core in an ASIC"
- Training and self-organized gain in experience concerning software development for reconfigurable systems

19.2.8 Intracom Telecom Solutions SA

INTRACOM currently implements telecommunications products based on DSP, FPGA, and ASIC platforms. In general, entry-level engineers have sufficient academic training on one (or more) of these platforms.

For emerging technologies, such as the one targeted by the MORPHEUS project, engineers would benefit from additional training in:

- H/W–S/W co-design principles and flows
- Reconfigurable H/W programming based on high-level languages (above RTL), such as C

So far, INTRACOM has addressed the main concepts of reconfiguration as part of research towards the direction of reconfigurable systems-on-chip. INTRACOM's researchers have been convinced that future company's products could benefit from the use of reconfigurable technology. The main concepts of reconfiguration have been presented to design teams through a series of introductory lectures, which constitute part of INTRACOM's internal seminars to telecom systems engineers. In that direction, a book entitled "System Level Design of Reconfigurable Systems-on-Chip" by Springer, has been published containing details on how future systems could benefit from the use of reconfigurable technologies.

For INTRACOM the MORPHEUS project, apart from the technological enhancement of the existing know-how on reconfigurable systems, will form a consolidated reconfigurable platform that could constitute the basis of future telecom products. In that respect, INTRACOM plans to offer continues and in-depth training of the design teams so as to become familiar with the technology and the methods available for the efficient design of reconfigurable systems. As far as INTRACOM is concerned, training requirements should rely on:

- Training on the main components that constitute the final MORPHEUS platform
- Training on the available tools and methodologies

Due to the traditional 'gap' that exists between hardware and software engineers, the effectiveness of the proposed training will come across a real challenge: to convince hardware designers to raise their level of abstraction (and as a result converging with software people), while in parallel train software engineers so as to take into account traditional/reconfigurable hardware aspects. In fact, this is a problem of hardware/software co-design, where the concepts of reconfiguration play a crucial role throughout system design. What is expected from this kind of training is to provide engineers with the required culture to design reconfigurable systems in a way that reconfiguration is taken into consideration at higher levels of abstraction.

Apart for the need to train INTRACOM's engineers to reconfigurable technologies, there is a strong need to provide a link to Greek universities so as to gradually include the concepts of reconfigurable system design in their regular courses. In this way, reconfigurable design culture will become part of young engineers' qualification, in a way similar to the training they currently have on hardware or software design methods and tools.

19.2.9 ACE

ACE Associated Compiler Experts are the vendor/provider of the professional compiler development system called CoSy. As such, ACE provides training facilities related to best practice compiler development with the CoSy system. Apart from the many industrial clients, CoSy is broadly used in the Academic world for research and development of compilers and compilation techniques. It would be beneficial for ACE, as well as the chip and compiler tools industry, if academia disseminate their CoSy experiences for reconfigurable and parallel architectures. As a result, the compiler development community will be encouraged to use CoSy as a professional and cost effective compiler platform.

19.3 Course Planning and Specification

The analysis of the training needs performed the segmentation of the training offer into three main classes:

1. Training for HW-Designers/Engineers
2. Training for SW-Designers/Engineers
3. Training for both HW and SW Designers/Engineers
4. Training for MORPHEUS end users

These main classes were now sub-segmented into training courses with the required content survey with the industry partners. In Table 19.1, the training requirements were bundled and bounded to specific courses. In the first column a possible course title is noted. The second column describes the content of the course provided within the MORPHEUS project.

Table 19.1 MORPHEUS course and training

Training for HW-Designers/Engineers: Course Title	**Content of Course**
Coarse-grained reconfigurable architectures	• State of the art coarse-grained reconfigurable architectures • Structure of the different processing elements • Coarse-grained reconfigurable architectures used in MORPHEUS (detailed) • Algorithm mapping to parallel architecture • Exploitation of dynamic reconfiguration
HW/SW interaction and design testability	• Requirement analysis for software demands • Cost estimation for hardware requirements after analysis for software demands

(continued)

Table 19.1 (continued)

	• Hardware requirements and parameterization in relation to driver software • Modular and testable design methods • Design of tests suites in correlation to software requirements
Training for SW-Designers/Engineers: Course Title	**Content of Course**
Parallel computing on reconfigurable architectures	• Basic concept of FPGA architecture • Mode of use: Software and Hardware partitioning • Abstraction levels for reconfigurable system in MORPHEUS: Top down and bottom up approach • Profiling and native programming • Mapping of algorithms to reconfigurable architecture
HW/SW-communication in heterogeneous reconfigurable SoC	• Data transfer in reconfigurable MORPHEUS SoC • Specification of external IP • Driver design for data transfer in reconfigurable architecture • Specification of hardware in relation to software-drivers • Multi-tasking and synchronisation of HW/SW partitioned systems
Training for Both HW and SW Designers/ Engineers: Course Title	**Content of Course**
Hardware–Software Co-design for reconfigurable SoCs	• Methods and algorithms for HW/SW partitioning • Graph Algorithms • Parallelisation of tasks • Basic knowledge of HDL coding • Bus-standards and Network-on-Chip • High level modelling of complex SoC architectures
Methodology of reconfiguration in MORPHEUS architecture – trade-off and benefits	• Training with MORPHEUS architectures • Power, area, performance trade off in heterogeneous reconfigurable SoCs • Cost awareness by exploiting reconfiguration (in terms of power, performance and area) • Concepts and methodology of reconfiguration
Simulation and analysis of heterogeneous SoC	• Memory usage in heterogeneous SoC architecture • Example: DMA setup and buffer handling • Platform synthesis and analysis of MORPHEUS architecture

(continued)

Table 19.1 (continued)

Applications for MORPHEUS architecture and project planning	• Telecommunication standards • Video data processing • Project handling: How to steer complex projects
Training for MORPHEUS End Users: Course Title	**Content of Course**
Application notes with MORPHEUS reconfigurable SoC	• Introduction to application scenarios • Introduction of toolchain for MORPHEUS • Description of flexibility through reconfiguration • Reconfiguration paradigm • Exemplarily: Integration of one example application on MORPHEUS platform

19.4 Training Events

19.4.1 AMWAS 2007

AMWAS 2007 (ÆTHER – MORPHEUS Workshop and Autumn School), "From Reconfigurable to Self-Adaptive Computing" 8–11 October 2007, Paris, France, was a common event organized between the ÆTHER and the MORPHEUS projects. The two projects have many points of common interest: in fact MORPHEUS is focusing on dynamic reconfigurable computing, whereas ÆTHER is studying self adaptive computing systems. Both projects are grounded on computer architecture and software tools. The two projects have in common a set of participants: Thales, CEA-LIST, UK and INTRACOM are in fact participating to both the projects.

The event was organized in two parts: the autumn school on reconfigurable computing and the autumn workshop on reconfigurable computing. The school was more focused on dissemination and training issues with a participation of 33 people, mainly students from the university partners of the two projects. The program of the school included, after the introduction from the coordinators G. Edelin and C. Gamrat, 11 presentations: 5 from MORPHEUS, 5 from ÆTHER and 1 common to both projects. The School was aimed at raising interest among students about a possible involvement in the projects or in possible further developments on reconfigurable computing. Table 19.1 contains a summary of the contents of the Autumn School; it reports only the parts about MORPHEUS.

Though emphasis was on innovative techniques developed by the AETHER and MORPHEUS research communities, the AMWAS program has not been restricted to the AETHER and MORPHEUS approach but it provided as well a comparative analysis of other proposals and experiences on reconfigurable and self-adaptive computing solutions, from the hardware level to the application level. Researchers and experts working in other EU-funded research projects participated, lecturing on each sub-theme. Various points of view have been presented in after-lecture discussion hours and in focused seminar sessions.

The main theme of the AMWAS'07 deals with the broad subject of "Self-adaptive dynamically reconfigurable Computing Architectures", focusing more specifically on the problem of self-adaptation at all levels within a system, from hardware architecture to application layer. Knowledge to be provided by the AMWAS refers to "Computing with Adaptive Hardware: Architecture and Software."

The topics were divided into several sub-themes, which were presented from the viewpoint of the various disciplines involved in the main AETHER and MORPHEUS subprojects. The program included the following sub-topics:

- Adaptive systems and self-adaptive systems
- Dynamic reconfiguration and self-adaptivity
- Parallelization and allocation of an application onto reconfigurable architectures

19.4.2 AMWAS 2008

The Second AETHER – MORPHEUS Workshop- Autumn School "From Reconfigurable to Self-Adaptive Computing" (AMWAS'08) was organized in Lugano, Switzerland from October 7–9, 2008. The innovative "School/Workshop" format has been chosen as second time to provide participants with both grounding on a new, challenging scientific area and with exposure to research results and proposals. The AMWAS'08 constituted a meeting-point for researchers and graduate students, interested in innovative next-generation computing architectures.

The event has been organized in a one and a half day school as well as a one and a half day workshop. Both events were focused on self-adaptive/dynamically reconfigurable computing architectures as in the year before.

19.5 Conclusions

After the assessment phase it becomes clear, that education needs to bridge the software and hardware world to provide well skilled engineers for the future MORPHEUS architecture. The paradigm shifts from pure CPU based embedded systems to heterogeneous architectures including coarse and fine grained reconfigurable architectures needs to be introduced in an early phase of study but should also be provided for well established engineers. The design space which is offered by the MORPHEUS architecture can only be handled if a deep understanding of the hardware parts is trained in order to exploit the dynamics while run-time of this future architecture. Certainly the toolset helps to distribute the tasks efficiently to the hardware but engineers needs to have an overview what happens if applications have to be implemented on the heterogeneous chip.

This basic knowledge will be transferred with improved and novel courses provided by the academic partners MORPHEUS partner in deep cooperation with the industry partners who can provide the necessary material of their provided architecture or tools. Only with this process, a common understanding about future dynamic hardware architectures can be established and managed.

References

1. R. Hartenstein, The Changing Role of Computer Architecture Education within CS Curricula, Workshop on Computer Architecture Education, Munich, Germany, June 19–23, 2004.
2. M. Hübner and J. Becker, Tutorial on Macro Design for Dynamic and Partially Reconfigurable Systems, First RC-Education Conference, 2006, Karlsruhe, Germany.
3. Reconfigurable Computing Workshop: A New Computing Paradigm. Impact on Thales Products: A Perspective, 18/09/03, Thales TRT, Domaine de Corbeville, Orsay, France.
4. Seminar on New Perspective in Parallel Processor Architecture, 30/06/04, Thales TRT, Domaine de Corbeville, Orsay, France (seminar including most presentations on reconfigurable computing).
5. V. Baumgarte, et al., PACT XPP – A self-reconfigurable data processing architecture. *Journal of Supercomputing*, 26(2):167–184, September 2003.
6. M2000's Spherical FPGA Cores, Max Baron, MicroProcessor Report, December 2004.
7. F. Campi, R. Canegallo, A. Cappelli, R. Guerrieri, A. La Rosa, L. Lavagno, A. Lodi, C. Passerone, M. Toma, A reconfigurable processor architecture and software development environment for embedded systems, Tenth IEEE Reconfigurable Architectures Workshop (RAW'03), Nice, France, April 22, 2003.

Chapter 20
Dissemination of MORPHEUS Results

Spreading the Knowledge Developed in the Project

Alberto Rosti

Abstract This chapter summarizes the dissemination strategy of the MORPHEUS project. It defines the target groups addressed by the dissemination activity and the actions planned for them. It comments about the media that have been used to promote the project. Finally it lists the main dissemination events and the publications.

Keywords Dissemination · publications · workshops · conferences · European projects

20.1 Introduction

Strong dissemination is decisive for the project's success. The final aim of research is the development of business opportunities; therefore the most important goal is bridging the gap between research and marketing. The MORPHEUS project has been looking for the most appropriate channels to spread the developed knowledge to the benefit of the widest research community.

20.2 Target Groups

Dissemination occurs across all partners to achieve consistent passing of message and value. In a dissemination plan the project partners identified the target groups and the actions to address them. The target groups of the dissemination are listed in Table 20.1 which describes their composition as well as the motivation to address them and the used means.

A. Rosti (✉)
STMicroelectronics, Italy
alberto.rosti@st.com

N. Voros et al. (eds.), *Dynamic System Reconfiguration in Heterogeneous Platforms*, Lecture Notes in Electrical Engineering 40,

Table 20.1 Target groups for dissemination

Target	Composition	Motivation	Means
Project partners	The partners of the MORPHEUS consortium	Foster any possibility of collaboration	– Project deliverables – Joint work-groups
Designer Community	The designer community in the business units of the consortium	Rise awareness about the new technology applied to design	– Workshops – Training – Demonstrations
European Community	Executives from the European Community	Monitor the dissemination activity	– Project deliverables – Press release
Other projects	Participants to other European research projects	Sharing knowledge, setting-up common practices	– Common events
Scientific Community	The wider scientific community out of the consortium	Enhance the state of the art in reconfigurable computing domain	– Conferences – Journals – Demonstrations – Book – Web site
University students	Participants to university advanced courses	Improve university courses	– Training – Summer school – Book
Industrial Community	The wider industrial community out of the consortium	Rise general awareness about RC technology	– Press release – Web site – Flyer

The project partners are both authors and recipients of the dissemination activity, they foster any opportunity to improve their collaboration.

The designer community inside the consortium is another important target; they are mainly recipients of the dissemination activity.

The European Community through the Project Officer and the Reviewers is addressed for monitoring the dissemination activity checking if the project is on track.

The participants to a set of European projects connected to MORPHEUS are addressed by specific actions planned for them.

The overall scientific community in the fields of Electronic Design Automation and Embedded Software is a natural target of our research activity.

Students from the universities are another important target group, they will be the designers of the future; thus it is very important to raise their interest about the possibilities offered by dynamically reconfigurable computing technology and advanced design methodology to solve real design problems. Students are involved mainly by training and by special events organized for them.

The last target group is the wider industrial community, to promote a more advanced European industry.

20.3 Dissemination Media

A set of complementary media is used for dissemination. It includes the press releases, the project flyer, the MORPHEUS book and the public web site; all of them will be described in this section.

20.3.1 Press Releases

A press release has been issued following an advice from the Project Officer and the Reviewers from the European Community. It is a formal document which defines the position of the consortium in the industrial community. It introduces the consortium and presents the motivations, the main technical challenges and the goals of the project to the European Community and to the industrial community.

The press release was issued in Paris by Thales on March 26, 2007; it is available on the MORPHEUS public web site at http://www.morpheus-ist.org/pages/pressrelease.htm.

Some feedback is available from the feast forum (feast is a Forum for European-Australian Science and Technology cooperation) reporting about "French Science and Technology Fortnight" (May 22, 2007). In this forum the activity of MORPHEUS was reported among the most interesting research topics, the project was presented as an initiative carried on by 18 partners from 6 European countries, including universities, small enterprises and large industries, aimed at engineering dynamically reconfigurable platforms, with the development of a proper design toolchain. The report was also mentioning the availability of the first project results.

A second press release is planned at the project completion.

20.3.2 Flyer

The project flyer is a brochure that illustrates the MORPHEUS project to the wider industrial community interested in dynamically reconfigurable computing. It introduces the project providing a set of basic information and contact references. The flyer illustrates MORPHEUS as a solution that combines the advantages of general purpose processors, FPGAs and ASICs in a single chip environment. The MORPHEUS platform is programmable like a general purpose processor, optimal like an ASIC and flexible like an FPGA. It offers a cost-effective, flexible and high performance "Domain focused Platform" which combines the advantages of existing embedded computing solutions, while avoiding their drawbacks.

The flyer was used for dissemination at the ECSI workshop 2007, at the AWMAS 2007 in Paris and at AMWAS 2008 in Lugano. The flyer has been presented at DATE in 2007 and in 2008 where its contents have been summarized in a poster. Reading the flyer people appreciated mostly the completeness of our design flow that entails an application-driven full path to silicon implementation defining the architecture, providing a tool-set for Reconfigurable Computers design and actually building the final silicon implementation.

20.3.3 MORPHEUS Book

This book has been inserted in our dissemination plan to present all the project achievements in a reader friendly way. From this book we expect a more systematic dissemination of the basic knowledge and the state of the art about dynamically reconfigurable computing. We also intend to provide information about the role of the MORPHEUS project in the domain and to provide details about the innovative points, demonstrating how it contributes to improve the state of the art.

20.3.4 Public Web Site

The project web site (Fig. 20.1) is the main communication channel out of the project. The web site describes the MORPHEUS consortium, it contains documentation and publishable abstracts of deliverables, it provides training material and publishes news about the main events, it finally proposes contacts to the visitors with the project partners. The site is hosted at http://www.morpheus-ist.org/ by ARCES.

This public web site aims at becoming a reference site about dynamically reconfigurable computing. Its contents are updated continuously throughout the whole duration of the project.

The data about accesses to the web site from January 1, 2008 until November 30, 2008 are summarized in Table 20.2.

20.4 Dissemination Activities

20.4.1 Workshops and Common Events

The UNICAD workshop is a major event organized every year in STMicroelectronics to present new developments in technologies and methodologies to the internal designer community and to the company's strategic customers. In 2006 the MORPHEUS team had the possibility to produce a presentation about the "Configurable data-path PiCoGA" demonstrating the technology. The feedback indicated interest in the evaluation of PiCoGA/DREAM and also of other

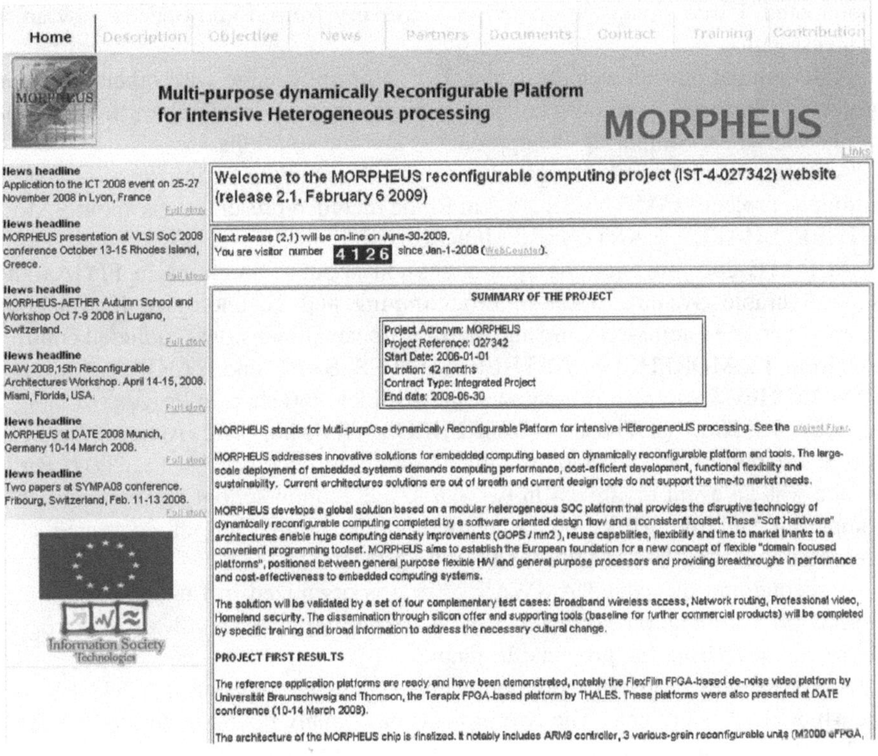

Fig. 20.1 The MORPHEUS web site

Table 20.2 Web usage statistics

Visits	Unique Visitors	First Time Visitors	Returning Visitors	Average Visits/Months
2,901	2,338	1,880	458	264

reconfigurable processing elements that are proposed within MORPHEUS such as FlexEOS and XPP-III.

The ECSI Workshop in Paris on January 18, 2007 was covering industry needs and requirements for current and future application scenarios of reconfigurable systems. MORPHEUS participated with four presentations about: "Dynamic Reconfigurable Hardware – Tools and Architectures for SoC", "Application programming design flow for the MORPHEUS dynamically reconfigurable platform", "Reconfigurable Computing Needs from an Industrial Perspective", and "Applications for reconfigurable systems-on-chip in telecommunication networks". This workshop gave us the opportunity to attract the attention about the needs for reconfigurable computing in industry, to establish the application-driven nature of the MORPHEUS project and to focus on the need of proper tools enabling design flow on reconfigurable architectures. MORPHEUS presentation was very well appreciated. Discussions with attendees were very fruitful and opened new areas of investigation.

A dissemination achievement is the sharing of knowledge with other European projects. This activity allows collecting feedback from researchers involved in similar activities, comparing the approaches and the solutions.

A set of events has been organized to promote sharing of information among European projects. CASTNESS 2007 in Rome included presentations about SARC, ÆTHER, SHAPES, HARTES and MORPHEUS.

At DATE 2007 in Nice the special session about "Directions in FPGAs and Reconfigurable Systems: Design, Programming and Technologies for adaptive heterogeneous Systems-on-Chip and their European Dimensions" included contributes from 4S, MORPHEUS, AETHER, SHAPES, SARC and HARTES.

MORPHEUS was also presented at the 4S final workshop in Prague.

AMWAS 2007 (ÆTHER – MORPHEUS Workshop and Autumn School), "From Reconfigurable to Self-Adaptive Computing" October 8–11, 2007, Paris, France was an event organized in two parts: the autumn school on reconfigurable computing and the workshop on reconfigurable computing. It contained also presentations from the SHAPES and SARC projects.

A second edition named AMWAS 2008 was organized in Lugano on October 7–9, 2008 focusing on software aspects, exploitation of hardware reconfigurability and preparing the project aftermath.

AMWAS workshops are common events organized between the ÆTHER and the MORPHEUS projects. The two projects have many points in common: in fact MORPHEUS is focusing on dynamic reconfigurable computing, whereas ÆTHER is studying self adaptive computing systems, both the projects are grounded in computer architecture and software tools. The two projects have in common a set of participants: Thales, CEA-LIST, UK and INTRACOM. Table 20.3 summarizes the European projects presented during the AMWAS workshops.

20.4.2 Demonstrations

At the Unicad Workshop 2006 STMicroelectronics demonstrated the configurable data-path PiCoGA/DREAM.

Table 20.3 Related European projects

ÆTHER	The project deals with Self-Adaptive Embedded Technologies for Pervasive Computing Architectures
SHAPES	The SHAPES project deals with tiled architectures for multimedia applications
4S	The projects designs Flexible heterogeneous (mixed Signal) platform
SARC	SARC is an integrated project concerned with long term research in advanced computer architectures
HARTES	It is aimed at bridging the gap between SW and architecture

At the university boot of DATE 2007 the University of Delf and ACE demonstrated the CoSy compiler based on MOLEN in a dedicated exhibition stand focused at presentation and demonstration of academic compiler research projects.

At DATE 2008 the University of Braunschweig presented the Flexim Real-time Digital Film Processing showing a demonstration of a board implementing FPGA-based reconfigurable platform. The University of Karlsruhe presented a demonstration about the MORPHEUS integrated toolset showing the design flow to map execution of a FFT on the PiCoGA/DREAM controlled by the ARM. The University of Chemnitz demonstrated the design specification with SpecScribe/SpecEdit.

20.4.3 Publications

The MORPHEUS project is characterized by a large publication activity which includes papers in conference proceedings and journals. It covers all the aspects of the project, from the enabling technologies furnished to MORPHEUS by the partners to the project results. We analyze those publications shortly grouping them according to the criteria that inspired them.

The purpose of introducing the reconfigurable computing technology is accomplished by a set of papers from the University of Karlsruhe [1,2]. They explore how to use FPGA for dynamic and partial real-time reconfiguration.

Alcatel-Lucent presented a set of papers [3–7] which propose to extend formal verification to the software domain deriving automatically the test from a formal system specification in the ADeVA language, applying model checking. These techniques are used in the specification of the applications in the MORPHEUS project.

A set of publications [8–11] outlines that MORPHEUS is the first project which provides a set of high level tools to program different kinds of reconfigurable units assembled on a system on chip. The toolset spans many issues: compilation, the MOLEN paradigm, RTOS dynamic scheduling of reconfiguration and spatial design of the accelerated functions on reconfigurable units.

Papers [12,13] deal with the PiCoGA/DREAM reconfigurable accelerator. Various applications on the target MORPHEUS platform are treated in papers [14–18]. Paper [19] introduces the network on chip. Finally papers [20–22] deal with the MORPHEUS project in all its aspects reporting about its status of advancement at the time of their publication.

20.5 Conclusions

The dissemination activity within the MORPHEUS project has been carried out since the early phases of the project (from year 2006) according to the goal described in the introduction of this chapter about filling the gap between research and marketing. The project defined a dissemination plan to identify the target groups and the most appropriate activities and means to address them.

In a first phase of the project the dissemination activities were mainly aimed at raising the awareness about the potential benefits of the reconfigurable computing technologies and at documenting the IPs that every partner was bringing into the project. In a second phase the dissemination activity was more focused on the results coming from the project.

The utmost attention has always been paid to the collection of feedbacks that were used to properly adjust the project trajectory.

References

1. M. Hübner and J. Becker, Tutorial on Macro Design for Dynamic and Partial Reconfigurable Systems, *Proceedings of 1st International Workshop on Reconfigurable Computing Education*, March 1, 2006, Karlsruhe, Germany.
2. M. Hübner and J. Becker, Exploiting Dynamic and Partial Reconfiguration for FPGAs – Toolflow, Architecture and System Integration, *Proceedings of 19th SBCCI Symposium on Integrated Circuits and Systems Design*, September 2006, Ouro Preto, Brazil.
3. A. Schneider, T. Blum, T. Renner, U. Heinkel, J. Knäblein, and R. Zavala, Formal Verification of Abstract System and Protocol Specification, *IEEE/NASA Software Engineering Workshop (SEW)*, April 2006, Columbia, MD, USA.
4. A. Schneider, G. Bunin, C. Haubelt, and U. Heinkel, Automatic Test Generation with Model Checking Techniques, *Proceedings of International Conference on Quality Engineering in Software Technology (CONQUEST)*, September 27, 2006, Berlin, Germany.
5. G. Bunin, A. Schneider, C. Haubelt, S. Gossens, J. Langer, and U. Heinkel, Automatic Test Case Generation with NuSMV, *MOTES06 Workshop Model-Based Testing*, October 2006, Dresden, Germany.
6. A. Schneider, S. Walter, J. Langer, and U. Heinkel, Automatic Visualization of Abstract System Specifications, *International Conference on Quality Software QSIC*, October 2006, Beijing, China.
7. A. Schneider and J. Langer, Generation of Test Automation Code with Model Checking, *12th Software and Systems Quality Conferences (SQS)*, April 27, 2007, Dusseldorf, Germany.
8. G. Edelin, P. Bonnot, W. Gouja,K. Bertels, F. Thoma, A. Schneider, J. Knäblein, B. Pottier, and J.C. Le Lann, A Programming Toolset Enabling Exploitation of Reconfiguration for Increased Flexibility in Future System-on-Chips, *DATE 2007*, April 16–20, 2007, Nice, France.
9. B. Pottier, J. Boukhobza, and T. Goubier, An Integrated Platform for Heterogeneous Reconfigurable Computing, *The International Conference on Engineering of Reconfigurable Systems and Algorithms (ERSA)*, June 25–28, 2007, Las Vegas, NV, USA.
10. J.C. LeLann, B. Pottier, M. Godet, and R. Keryell, Loosely Coupled Accelerators for Reconfigurable Systems on Chips, *IEEE International Parallel & Distributed Processing Symposium*, March 2007, California, USA.
11. E. Moscu Panainte, K.L.M. Bertels, and S. Vassiliadis, The Molen Compiler for Reconfigurable Processors, *ACM Transactions in Embedded Computing Systems (TECS)*, February 2007, Volume 6.
12. A. Lodi, C. Mucci, M. Bocchi, A. Cappelli, M. De Dominici, and L. Ciccarelli, A Multi Context Pipelined Array for Embedded Systems, *FPL 2006*, August 2006, Madrid, Spain.
13. F. Campi, A. Deledda, M. Pizzotti, L. Ciccarelli, C. Mucci, A. Lodi, A. Vitkovski, and L. Vanzolini, A Dynamically Adaptive DSP for Heterogeneous Reconfigurable Platforms, *Proceedings of DATE conference*, April 16–20, 2007, Nice, France.

14. S. Heithecker, A. do Carmo Lucas, and R. Ernst, A High-End Real-Time Digital Film Processing Reconfigurable Platform, *EURASIP Journal on Embedded Systems Volume 2007*, December 2006.
15. A. Schneider, J. Knäblein, B. Müller, M. Putsche, S. Goller, U. Pross, and U. Heinkel, Ethernet Based In-Service Reconfiguration of SoCs in Telecommunication Networks, *4th Workshop on Dynamically Reconfigurable Systems (DRS)*, March 12–15, 2007, Zurich, Switzerland.
16. C. Mucci, L. Vanzolini, A. Deledda, F. Campi, and G. Gaillat, Intelligent Cameras and Embedded Reconfigurable Computing: A Case Study on Motion Detection, *International Symposium on System-on-Chip*, November 19–21, 2007, Tampere, Finland.
17. C. Mucci, L. Vanzolini, F. Campi, A. Lodi, A. Deledda, M. Toma, and R. Guerrieri, Implementation of AES/Rijndael on a Dynamically Reconfigurable Architecture, *Proceedings of DATE Conference*, April 16–20, 2007, Nice, France.
18. C. Mucci, L. Vanzolini, I. Mirimin, D. Gazzola, A. Deledda, S. Goller, J. Knaeblein, A. Schneider, L. Ciccarelli, and F. Campi, Implementation of Parallel LFSR-Based Applications on an Adaptive DSP Featuring a Pipelined Configurable Gate Array, *Proceedings of DATE Conference*, April 16–20, 2008, Munich, Germany.
19. A. Deledda, C. Mucci, A. Vitkovski, M. Kuehnle, F. Ries, M. Hübner, J. Becker, P. Bonnot, A. Grasset, P. Millet, M. Coppola, L. Pieralisi, R. Locatelli, G. Maruccia, F. Campi, and T. DeMarco, Design of A HW/SW Communication Infrastructure for a Heterogeneous Reconfigurable Processor, *Proceedings of DATE Conference*, April 16–20, 2008, Munich, Germany.
20. M. Kühnle, F. Thoma, M. Hübner, and J. Becker, University Booth: MORPHEUS Multi-Purpose Dynamically Reconfigurable Platform for Intensive Heterogeneous Processing, *DATE Conference*, April 16–20, 2007, Nice, France.
21. F. Thoma, M. Kühnle, P. Bonnot, and S. Goller, MORPHEUS: Heterogeneous Reconfigurable Computing, *FPL 2007 17th International Conference on Field Programmable Logic and Applications*, August 27–29, 2007, Amsterdam, Netherlands.
22. A. Rosti, P. Bonnot, S. Perissakis, K. Potamianos, and W. Putzke-Röming, MORPHEUS – Heterogeneous Reconfigurable SOC, *VLSI SoC 2008*, October 13–15, 2008, Rhodes Island, Greece.

Chapter 21
Exploitation from the MORPHEUS Project

Perspectives of Exploitation About the Project Results

Alberto Rosti

Abstract This chapter analyzes briefly the perspective of exploitation of the results developed in the MORPHEUS project. After defining our concept of exploitation this chapter outlines who are the actors of the exploitation, what is the matter of the exploitation and the target of the exploitation. All the considerations that we make are organized according to the classes of partners that compose the consortium: large industries, small medium enterprises and universities.

Keywords Exploitation · marketing

21.1 Introduction

The consortium agrees that the exploitation within MORPHEUS is globally intended as the strategy to make the project used and developed further.

Up to now, every large industrial partner has prepared the exploitation plan according to its particular vision and focus. Those exploitation plans have been defined contacting the business units (internal or external to the company, depending on the organizations). The exploitations perspectives are various: there are partners which are promoting partial exploitation of developments carried out within the MORPHEUS project, aiming at proving the concept of reconfigurable computing for further products.

SMEs are generally interested in sustaining the marketing road map of their products that they bring as contribution to MORPHEUS, leveraging on the go-to-silicon path provided by the project, enhancing their tool flows, gaining visibility, sustaining the growth of the company.

The academic partners are more focused on training, dissemination, further research through new projects and exploitation of patents.

A. Rosti (✉)
STMicroelectronics, Italy
alberto.rosti@st.com

N. Voros et al. (eds.), *Dynamic System Reconfiguration in Heterogeneous Platforms*, Lecture Notes in Electrical Engineering 40,

21.2 Who is Performing Exploitation

About the actors of exploitation we must say that every organization in the consortium is actively participating to the exploitation. The project consortium is made of large industries, small medium enterprises and universities. So the project has many different possibilities for exploitation because each partner is interested in exploiting different aspects of the project. Our experience in the project demonstrated that experts in marketing provide a more realistic feeling about the real needs for exploitation; the best thing is to have full collaboration from marketing and technical people.

21.3 What is the Matter of Exploitation

Some large companies are aimed at obtaining proof-of-concept out of the project for further developments. For instance they are willing to evaluate the possibility to introduce a specialized reconfigurable platform into the next generation of product lines.

Large industrial partners are mainly interested in exploitation internally to their companies. They mainly want to assess the performances of reconfigurable IPs in different application domains.

SMEs want to increase the size of the company, develop new partnerships, sustain their product roadmap, sustain new projects and increase their visibility. The participation to the project is a marketing achievement.

Universities are using the results of the project to improve curricula of regular courses and to set up exercise/laboratory sessions, improving the quality of engineering education. An improved curriculum of study is already an outcome of the project.

Many partners are participating to the project in order to develop, enhance and promote their tools for reconfigurable computing.

In all the cases from the project the partners are willing to acquire technology and know-how that they will use for further development of architectures, tools and applications.

In some case the matter of interest for exploitation are directly the MORPHEUS chip and board.

In summary, the expected exploitation results cover the following three areas. The hardware architecture in all its components: mainly the configurable engines and the communication infrastructure. A second exploitation area is in the tool chain, which after refinement and integration can be a base of a new general reconfigurable design methodology. This area of exploitation right now is more in the ambitions of some university partners and not yet enough mature from an industrial point of view. The third area is in the applications, relying on the results of the four demonstrator projects; particular care is paid to the mapping of the application onto the reconfigurable platform. Some partners working on the applications are looking for an architecture similar to an ASIC with embedded reconfiguration engine to provide enough flexibility and throughput for their demanding application.

21.4 Target of Exploitation

The project partners have different exploitation possibilities depending on their main field of activity.

Large industrial partners are mainly interested in internal exploitation; they foster innovation through contacts with their business units. Some large companies are aimed at obtaining proof-of-concept out of the MORPHEUS project for further developments.

SMEs are contacting new customers, increasing the size of the company, developing new partnerships, sustaining their product roadmap, sustaining new projects and are aimed to obtain increased visibility.

Universities are using the results of the project improving curricula of regular courses and to set up exercise/laboratory sessions, improving the quality of engineering education. They are also giving rise to new projects and exploiting knowledge through patents.

ST's exploitation strategy involves the collaboration of internal business units such as the wireline division or groups that are performing evaluating the platform with applications. Another important outcome of the MORPHEUS partnership is the experimentation of the PiCoGA module for Thales telecommunication applications.

PACT presents the roadmap of its XPP-III architecture, the business model and plan are showing growing revenues from the XPP architecture rising from 2 M€ in 2007 to 9 M€ in 2008 and 30 M€ in 2009. They present a market analysis by target markets. They describe customers' needs that want to replace ASIC design with flexible programmable units. PACT has also a university program to promote its platform. PACT participates to MORPHEUS to increase visibility. The MORPHEUS project will increase the acceptance of reconfigurable computing solutions: a silicon implementation is mandatory for customer acceptance.

M2000 presents the exploitation about its eFPGA. Its inclusion in the MORPHEUS chip is a strong reference for M2000. Today's business model is to supply embedded FPGAs hard macros. It will be complemented by the direct sales of the new family of FPGA devices. The company is interested in attracting the Venture Capital firms. M2000 has grown from 19 people in Q3 2006 to 48 at the end of 2007 and more than 60 people in 2008. The target markets are presented for eFPGA and FPGA. M2000 lists various benefits from the participation to MORPHEUS such as: availability of a complete demonstrator platform (including hardware, software and applications), increased visibility on the market and improved industrial relations.

CBLUE wants to work closely with academic and industrial partners to enhance its tools. CBLUE has a commercial interest in reconfigurable architectures as this area will be a significant long term market opportunity. CBLUE believes that the availability of efficient tools for mapping software applications onto reconfigurable fabrics is a vital element that will ultimately produce significant end-product and designer benefits. The technologies being developed within the consortium are highly complementary to CBLUE's existing tools. The MORPHEUS project allows CBLUE to develop a high level of interoperability with these other tools and to make significant improvements to its Cascade tool to improve its targeting of reconfigurable architectures. CBLUE believes that this will widen the appeal of its own offering and will validate and grow the overall market opportunity for European reconfigurable fabric and tools vendors.

ACE in the MORPHEUS project wants to expand its knowledge and expertise in the field of compilation for reconfigurable processor architectures and parallel processor architectures. ACE wants to extend its existing CoSy compiler development system product with generic support for reconfigurable architectures and parallel architectures.

TRT develops links with Thales business units for the exploitation of the MORPHEUS results. The role of TRT is to provide competitive advantages to Thales through research studies. TRT organizes Techno Days to disseminate new technologies to Thales business units. The exploitation perspectives of the MORPHEUS project are mainly for proposing advanced reconfiguration concepts. The MORPHEUS platform will offer a state-of-the-art sustainable reconfigurable solution for Thales's computing applications.

DTO researches technologies and acts as strategic market consultant for Thomson BUs beyond 2 years time frame. DTO choose the film grain removal application, from Digital Film Technology GmbH[1] (DFT) to demonstrate the benefits of the MORPHEUS approach. The DFT market is characterized by very high end applications, low volumes, high prices, high data rates and sophisticated solutions. DFT offers equipment of complete range of high-quality solutions for acquisition, conversion, post production, and delivery systems for multiple formats. Typical customers are film post production firms. Higher scanning speed and resolution, better image stability, faster image processing solutions, automated workflows are main challenges for future products. MORPHEUS can enable the development of very flexible image processing architectures which can be reprogrammed for many different applications. Using such technologies would enable DFT to address many different product niches with the same platform, which was only possible from software solutions.

ICOM analyzes the market drivers leading to a wireless instead of a wired solution. The application targeted by INTRACOM S.A. Telecom Solutions is the emerging IEEE 802.16j broadband mobile wireless standard. ICOM has a wide portfolio of solutions including wireless access network infrastructure and is poised to promote the use of broadband wireless access, in full co-operation to wire-line broadband access networks. At present, the company's products are sold in over 50 countries, including the Eastern Europe region, the Americas and the Middle East. For many products, reconfiguration is a key factor for commercial success. The IEEE 802.16j broadband mobile wireless system is a demanding application which requires dynamic reconfiguration. Therefore, ICOM will prove the capabilities of MORPHEUS on an application that is beyond the state-of-the-art. The main

[1] The Digital Film Technology GmbH is the successor of the former Grass Valley Germany GmbH which was a Thomson company.

goal of ICOM is to use the IEEE 802.16j system as a proof of concept for applying MORPHEUS reconfiguration concepts. In parallel, ICOM envisages to apply these concepts in other wireless products of the same family as well.

ALCATEL-LUCENT presents its exploitation perspectives for reconfigurable systems-on-chip in telecommunication networks. The market trends indicate that there will be an evolution towards the "All-IP" network of the future where services drive technical innovation. Reconfigurable SoC can provide fundamental contribution to meet the user expectations facing the technical challenges. Reconfigurable computing will correspond to early market presence and shorten development cycle developing products before standards are stable. In the MORPHEUS project the contribution of ALCATEL-LUCENT mainly consists of the development of a concept for updating the hardware of installed network elements partially, a patent for this concept is currently pending. In phase 1 of the project, a demonstrator based on two FPGA evaluation boards has been build in order to proof the concept. In phase 2 the application was ported to the MORPHEUS platform. ALCATEL-LUCENT is not intending to use MORPHEUS chip as a product but only a demonstrator to prove a concept, moreover since the MORPHEUS chip will not be a commercial product, it will have to be adapted to the needs of the applications later on. ALCATEL-LUCENT does not expect to need all the parts of MORPHEUS chip. The MORPHEUS chip could then be scaled down to be adapted to the needs. The ALCATEL-LUCENT application is used to specify capabilities requirements for the MORPHEUS platform (power requirements, performances, flexibility, etc....).

TOSA intends to perform exploitation for handheld cameras, the MORPHEUS architecture appears to be a better candidate than FPGA since it brings low power consumption in addition to high processing power and strong reconfiguration capabilities. Hence, provided the MORPHEUS architecture confirms these expected benefits, TOSA plans to use such architecture for the handheld camera market.

UK will exploit the result of the project in particular about the on chip memory architecture, the SoC interconnect infrastructure (using and customizing the NoC) and the dynamic reconfiguration methodologies. The main exploitation activity performed by UK is in the training. The results of MORPHEUS project are used in a set of courses and a system-on-chip laboratory was started in summer 2006 dealing with reconfigurable computing issues. It is planned for practical introduction of the MORPHEUS platform when available. Remaining exploitation issues are related to the exploitation by other projects such as the started "Initiative reconfigurable supercomputing in KIT" and the possibility to reuse the MORPHEUS simulator platform in ÆTHER.

TUD in collaboration with ACE is mainly involved in the retargetable compilation tools for the target platform. One main exploitation result is the implementation in the CoSy compiler of the MOLEN programming paradigm which provides generic support for reconfigurable architectures. An additional exploitation result is the development of research contacts and collaborations with prominent researches in the reconfigurable computing domain in Europe.

CEA-LIST aims at the valorization of research projects results. Valorization is obtained through Publications, Patents, Prototypes, and Products. Reconfigurable architectures are considered as one key technology for embedded computing as they allow a good compromise between flexibility, processing power and consumption. MORPHEUS is the opportunity to extend research activity on reconfigurable architecture towards system management of computing resources. The MORPHEUS project will allow making very important progress in the area of reconfigurable computing. After the end of the project, the results will be used as starting point for more downstream projects and/or industrial collaboration contracts. They will also be integrated within CEA-LIST's internal embedded multi-processor computing platform as an asset for further research actions.

UBO is developing Master courses on software for embedded systems. Some of the developments achieved during MORPHEUS at UBO can be reused inside this course. It is the case for SoC high-level simulator that allows students to understand relationship between communication delays and bandwidth and computation circuit capabilities, as well as circulation of control information. It is also the case for synthesis and synthesis framework as shown from CDFG to M2000 technologies. UBO contributes to partner efforts to spread reconfigurable technologies in education (initiatives from UK is an example). A one time publication of the software will be done at the end of the project, as we cannot maintain a free software release with the current lab resources.

ARCES in collaboration with ST is mainly involved in the definition of the memory infrastructure and in the design of the overall interconnect. The achievements of the MORPHEUS project are used in three different university courses: "Digital Circuit Design", "Architecture for signal processing elaboration" and "Digital Electronics".

TUBS collaborates closely with DTO within the MORPHEUS project. The MORPHEUS project and its results are used in several lectures. The dissemination of the project results is an important task for the institute to raise interest in the MORPHEUS project and to promote the institute itself. The last important factor of the TUBS exploitation plan is the usage of project results for future research projects.

TUC will use MORPHEUS to extend its position in electrical engineering research in Saxony. MORPHEUS will help TUC to understand the reconfiguration algorithms provided by the FPGA vendors enabling TUC to extend and improve these tools.

21.5 Conclusions

A first exploitation area is the hardware architecture in all its components: mainly the configurable engines and the communication infrastructure.

A second exploitation area is the tool chain, which after refinement and integration can be a base of a new general reconfigurable design methodology.

The third area is in the applications, relying on the results of the four demonstrator projects. Particular care is paid to the mapping of the application onto the reconfigurable platform. Some partners working on the applications are looking for architecture similar to an ASIC with embedded reconfiguration engine to provide enough flexibility and throughput for their demanding application.

Chapter 22
Project Management

Hélène Gros

Abstract This chapter shows how the MORPHEUS EU project has been managed.

Keywords Management • project • monitoring • European Commission • Coordinator

22.1 Introduction

The objective of MORPHEUS management is to ensure that the work is performed properly at all levels and all times so as to achieve satisfactory the contract and to meet partners' expectations. The management of MORPHEUS was conducted by the Coordinator TRT, and ARTTIC, a company specialised in the management of R&D collaborative projects whose role is to facilitate the R&D monitoring, in accordance with the rules of the European Commission and with the Consortium Agreement.

Project management is implemented in the MORPHEUS project through the Work Package 1. It deals with technical/scientific, strategic and operational management. As research projects can deviate from their original purpose, it is therefore important that they are very well planned and that regular monitoring and communication is well set up to avoid losing control of the project. A regular monitoring of the budget consumption by the partners is also necessary.

22.2 MORPHEUS Management Structure

Good management means clearly identified responsibilities and rules, identified in the project organization. Therefore, three kinds of bodies have been set up within the project:

H. Gros (✉)
ARTTIC SAS, France
gros@arttic.fr

N. Voros et al. (eds.), *Dynamic System Reconfiguration in Heterogeneous Platforms*, Lecture Notes in Electrical Engineering 40,

- Decision bodies
  - The General Assembly (GA) which consists of a senior representative from each contractor and is chaired by the MORPHEUS Coordinator.
  - The Executive Board (EB) composed of the Coordinator and the WP leaders. The Executive Board has the overall technical management responsibility of the project and meets every 3 months.
- Advice and support teams
  - Executive Board Office (EBO) composed of ARTTIC. It brings support services such as financial control (annual cost statements, scheduling), administrative work (contracts, organization of project reviews, meetings, templates of MORPHEUS documents); help desk (support for MORPHEUS partners regarding administrative aspects), tools for communication and remote collaboration (including secured web platform).
  - Scientific and Technical Committee (STC) composed of recognized experts who provide Scientific and Technical advices to the Executive Board through technical review meetings or through some key deliverables review.
  - Exploitation Board (ExB), composed of partners of the Consortium whose aim is to give advices on how to enhance exploitation.
- Operational bodies
  - Work packages leaders. Their main responsibilities are to plan and manage the day-to-day work within its WP, and report it to the Coordinator (deliverables, WP reports …).

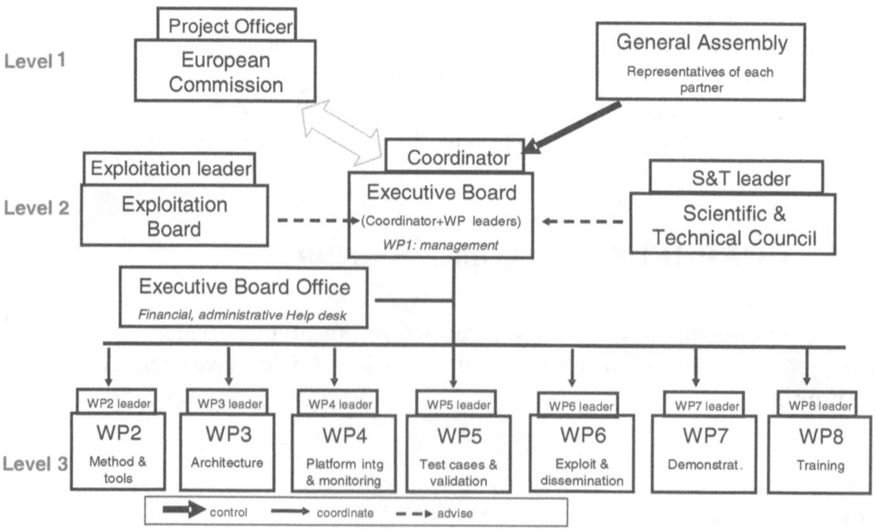

Fig. 22.1 MORPHEUS organizational structure

– Coordinator. The Coordinator is the intermediary between the consortium and the EC (supported by ARTTIC acting on his behalf) for all information related to the project. He is responsible for budget management.

Figure 22.1 presents an overview of the MORPHEUS organizational structure.

22.3 Monitoring the Project

22.3.1 Documents

The project management is based on a range of contractual documents. The main ones are the EC contract with its annexes (Description of work – DoW i.e. Annex 1, Annex 2, Annex 3, Forms A-B-C) and the Consortium Agreement. A Detailed Implementation Plan (DIP) describes the part of the contract which is updated every year containing the tasks, deliverables, milestones and efforts for the 18 coming month. Therefore, management includes the handling of the contractual EC requests to monitor the project: reports, review deliverables as well as additional internal reporting.

22.3.2 Deliverables

Deliverables are documents or prototypes showing major project results to allow the EC to follow the technical achievements of the project. Each significant element of the project should conclude with a deliverable as concrete output and evidence of work. Each deliverable tackles a specific subject, and has an owner responsible for the production of the document and coordination of eventually required partners' contributions. In order to plan and follow carefully the production of the deliverable, an internal "Deliverable Development Planning" has been set up. Moreover, both the form and the content of a deliverable are internally reviewed for quality improvement before sending the deliverable to the EC.

22.3.3 Reports

The consortium submits the following reports to the EC for each reporting period (at the end of each year of the project): (a) a Periodic Activity Report (PAR) containing an overview of the activities carried out by the consortium during that period, a description of progress toward the objectives of the project, a description of progress towards the milestones and deliverables foreseen, the identification of the problems encountered and corrective action taken. (b) A Periodic Management Report on that period including financial information (claim costs) from each contractor and in particular: a justification of the resources deployed by each contractor, linking them to activities implemented and justifying their necessity.

22.3.4 Indicators

A set of indicators have been implemented to monitor the project. The aim is to follow closely the progress and to provide consolidated information on the status of key-data (risks evolution, Person Month consumption, deliverables, milestones ...). The indicators are analyzed and updated regularly before being published on the Collaborative platform. In Fig. 22.2 we can see the milestone indicator.

22.3.5 Communication

22.3.5.1 Internal Communications

MORPHEUS partners use extensively electronic means to facilitate communication and exchange of information. Apart form the e-mails, the principal mean is the MORPHEUS collaborative platform which is a private web site dedicated to the project. It contains all information relative to the project (official documents, future meetings, minutes, news ...), documents archives and allows partners to work together, exchange information and work on documents such as project reports, deliverables...The web site is administrated by the EBO (ART).

22.3.5.2 Meetings

There are two kinds of meetings according to their location: physical meetings and conference calls. There are carried out in different occasions and for different scopes. The GA and EB meetings are periodic (every 3 months for the EB meetings) whereas the WP and cross WP meetings are planned by WP leaders

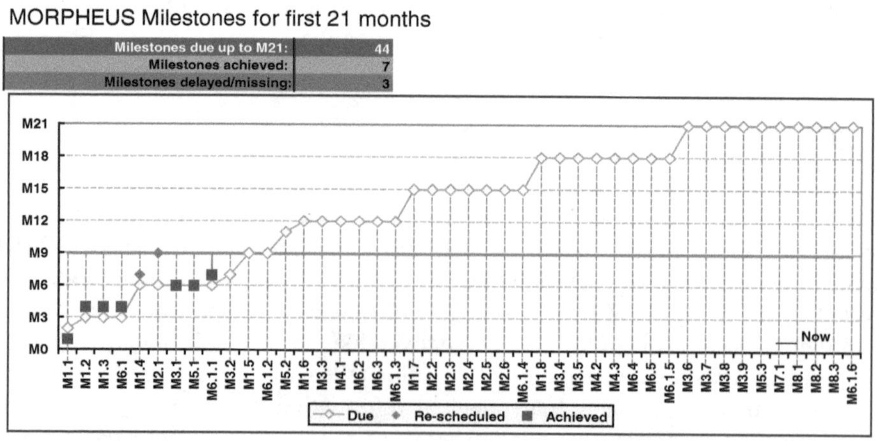

Fig. 22.2 The milestone indicator

when needed. The objectives, agenda and required contributions from the attendees are defined in advance (at least 2 weeks before the meetings). Minutes of the meeting are written by organizer and send to the participants within 2 weeks after the meeting. An Action list resulting from the meeting is also set up to follow closely the futures actions to be implemented in the coming days or months.

22.3.5.3 Publications

The EC strongly recommends the EU projects to disseminate their results. Before publishing information related to MORPHEUS, the partner must check that there is no conflict of interests with the other partners. A request for publication must be made to the GA. Once a text is agreed for publication, it will be stored in a dedicated space on the MORPHEUS collaborative platform for further access or reference.

22.4 Financial Management

The budget of the project should be followed carefully. Indeed, a regular monitoring of the consumption of their budget by the partners must be done to reallocate funding if necessary.

The global budget of the project which includes all eligible costs, is about 15,000,000€ with a funding from the European Commission of 8,240,000€. The financial rules for the participation in MORPHEUS are described in Annex 2 of the MORPHEUS contract and follow the FP6 EC financial guidelines. The amount of the EC financial contribution is determined by a number of factors including: the instrument, the activity and the cost model. Table 22.1 exhibits the maximum EC reimbursement rates of eligible costs.

Each year of the project, new advance payments are computed for the next 18 months period according to the new budget for the next 18 months, the approved costs claimed for the last 12 months and the previous advance payments. The money is transferred from the EC to the Coordinator after the approval

Table 22.1 Maximum EC reimbursement rates of eligible costs

Maximum Reimbursement Rates of Eligible Costs	RTD or Innovation Activities	Demonstration Activities	Training Activities	Management of the Consortium Activities
Integrated project	FC/FCF[a]: 50%; AC: 100%	FC/FCF: 35%; AC: 100%	FC/FCF: 100%; AC: 100%	100% (up to 7% of the contribution)

[a] Method of calculating indirect costs: FC = full costs with actual indirect costs; FCF = full costs with indirect flat rate costs; AC = additional costs with indirect flat rate costs.

of the cost claims, the annual reporting, the new DOW and DIP and the annual Review with the EC. According to the Consortium Agreement, the Coordinator organizes the distribution of the funding to the partners.

As conclusion, if project management requires an extensive work (information, coordination, financial management …) it also has a strategic dimension to ensure the quality of the project.

List of Acronyms

A/D	Analog to Digital
ÆTHER	Self-Adaptive Embedded Technology for Pervasive Computing Architectures
AG	Address Generator
ALU	Arithmetic and Logic Unit
AMBA	Advanced Microcontroller Bus Architecture
API	Application Programming Interface
ASIC	Application Specific Integrated Circuit
ATPG	Automatic Test Pattern Generation
BS	Base Station
CCG	Configuration Call Graph
CDFG	Control Data Flow Graph
CEB	Configuration Exchange Buffer
CMC	Central Memory Controller
COTS	Components of the Shelf
CPU	Central Processing Unit
CRC	Cyclic Redundancy Check
CSP	Communicating Sequential Process
D/A	Digital to Analog
DCT	Discrete Cosine Transform
DDR	Double Data Rate
DDRAM	Double Data Rate Synchronous DRAM
DEB	Data Exchange Buffer
DFG	Data Flow Graph
DMA	Direct Memory Access
DMU	Data Mover Unit
DNA	Direct Network Access
DSP	Digital Signal Processing
DWT	Discrete Wavelet Transformation
EDA	Electronic Design Automation
EDIF	Electronic Data Interchange Format
eFPGA	Embedded FPGA

ELF	Executable and Linkable Format
FF	Flip-Flop
FFT	Fast Fourier Transform
FIFO	First In First Out
FIR	Finite Impulse Response
FPGA	Field Programmable Gate Array
FPU	Floating Point Unit
FSM	Finite State Machine
GALS	Globally Asynchronous Locally Synchronous
GOPS	Giga Operation Per Second
GPL	GNU General Public License
GPP	General Purpose Processor
GUI	Graphical User Interface
HAL	Hardware Abstraction Layer
HDL	Hardware Description Language
HLL	High Level Language
HRE	Hardware Reconfigurable Engine
HW	Hardware
I/O	Input Output
ILP	Instruction Level Parallelism
IP	Intellectual Property: it defines any formalized reusable knowledge (hardware, software, or process)
IP	Internet Protocol
IPC	Instruction Per Cycle
ISA	Instruction Set Architecture
ISO	International Stardardization Organization
ISRC	Intelligent Services for Reconfigurable Computing
ISS	Instruction Set Simulator
KPN	Khan Process Network
LUT	Look Up Table
MAC	Media Access Control
MC	Motion Compensation
ME	Motion Estimation
MFC	Multi-Function Logic Cell
MIMO	Multiple Input Multiple Output
MORPHEUS	Multi-purpose dynamically Reconfigurable Platform for Intensive Heterogeneous Processing
MRS	Mobile Relay Station
MS	Mobile Station
MTT	Mode Transition Table
NML	Native Mapping Language
NoC	Network on Chip
OFDMA	Orthogonal Frequency Division Multiple Access
OTN	Optical Transport Network
PAE	Processing Array Element

PC	Personal Computer
PCI	Peripheral Component Interconnect
PCM	Prefetch Configuration Manager
PCMCIA	Personal Computer Memory Card International
PHY	Physical
PN	Petri Nets
POSIX	Portable Operating System Interface for uniX
PVM	Parallel Virtual Machine
RAM	Random Access Memory
RF	Radio Frequency
RISC	Reduced Instruction Set Computer
RLC	Reconfigurable Logic Cell
RS	Relay Station
RTL	Register Transfer Level
RTOS	Real Time Operating System
RTR	Run Time Reconfiguration
SAD	Sum of Absolute Differences
SAL	Symbolic Analysis Laboratory
SDK	Software Development Kit
SDRAM	Synchronous Dynamic Random Access Memory
SD-TV	Standard Definition TeleVision
SFU	Side Function Unit
SIMD	Single Instruction Multiple Data
SMV	Symbolic Model Verifier
SoC	System on Chip
SRAM	Static Random Access Memory
SW	Software
TCM	Tightly Coupled Memory
TDMA	Time Division Multiple Access
TIC	Test Interface Control
VLIW	Very Long Instruction Word
WCET	Worst Case Execution Time
WiMAX	Worldwide Interoperability for Microwave Access
XR	Exchange Register

Index

X